WORLD BANK TECHNICAL PAPER NUMBER 115

(Dam Safety and the Environment)

Guy Le Moigne, Shawki Barghouti,
and Hervé Plusquellec, editors

The World Bank
Washington, D.C.

Technical Papers are not formal publications of the World Bank, and are circulated to encourage discussion and comment and to communicate the results of the Bank's work quickly to the development community; citation and the use of these papers should take account of their provisional character. The findings, interpretations, and conclusions expressed in this paper are entirely those of the author(s) and should not be attributed in any manner to the World Bank, to its affiliated organizations, or to members of its Board of Executive Directors or the countries they represent. Any maps that accompany the text have been prepared solely for the convenience of readers; the designations and presentation of material in them do not imply the expression of any opinion whatsoever on the part of the World Bank, its affiliates, or its Board or member countries concerning the legal status of any country, territory, city, or area or of the authorities thereof or concerning the delimitation of its boundaries or its national affiliation.

Because of the informality and to present the results of research with the least possible delay, the typescript has not been prepared in accordance with the procedures appropriate to formal printed texts, and the World Bank accepts no responsibility for errors.

The complete backlist of publications from the World Bank is shown in the annual *Index of Publications*, which contains an alphabetical title list and indexes of subjects, authors, and countries and regions; it is of value principally to libraries and institutional purchasers. The latest edition is available free of charge from the Publications Sales Unit, Department F, The World Bank, 1818 H Street, N.W., Washington, D.C. 20433, U.S.A., or from Publications, The World Bank, 66, avenue d'Iéna, 75116 Paris, France.

Guy Le Moigne is senior adviser on water resources and agriculture in the World Bank's Agriculture and Rural Development Department, and Hervé Plusquellec is an irrigation engineering adviser in the same department. Shawki Barghouti is chief of the department's Agriculture Production and Services Division.

Library of Congress Cataloging-in-Publication Data

Dam safety and the environment / Guy Le Moigne, Shawki M. Barghouti,
  and Herve Plusquellec, editors.
      p.   cm. — (World Bank technical paper ; no. 115)
   Papers presented at the Workshop on Dam Safety and Environment
held in Washington, D.C., in April 1989.
   Includes bibliographical references.
   ISBN 0-8213-1438-6
   1. Dam safety—Congresses. 2. Dams—Environmental aspects—
Congresses. I. Le Moigne, Guy J.-M., 1932–   II. Barghouti, Shawki
M. III. Plusquellec, Herve. IV. Series.
TC540.D35 1990
363.3'493—dc20   89-78039
                                                              CIP

ABSTRACT

        This volume is a compendium of papers presented at the Seminar on
Dam Safety and the Environment held in Washington, D.C., in April 1989, and
sponsored by the World Bank. The main objective of the seminar was to
consolidate the experience of World Bank staff and of representatives from the
consulting industry and government agencies, in the design and management of
dam projects in order to improve their safety and lessen their environment
impact.

        The first part of the proceedings presents a historical and
geographic perspective on the construction of dams and a balanced overview of
the need for and benefits derived from dams, and of the increased concern for
their safety and the impact of dams on the environment. This section also
includes a presentation of the World Bank's role in dam construction.

        The second part focuses on (i) the experiences of the World Bank
in dam safety and the implementation of its policies; (ii) the experience of
developed and developing countries in the field of dam safety; (iii) design
for dam safety; and (iv) reassessing the safety of existing dams.

        The third part focuses on environmental aspects with special
reference to water quality and public health. It also presents a discussion
on the need to express environmental and social aspects in monetary terms in
the evaluation of dam projects.

        The seminar is a reflection of the continuation of the World Bank's
increasing interest in and concern with environmental and safety-related
aspects associated with the construction of dams.

CONTENTS

PREFACE

The World Bank, by providing funds to more than a hundred countries for over 400 projects involving dams, is the largest single funding agency of major dams in the world. Although its contribution representing less than 5 percent of the annual average number of dams built in the world is relatively modest, the World Bank has acquired wide experience and knowledge of the implementation of such projects.

Because of its prominent position in development, the World Bank has been the target of certain groups concerned with environmental aspects and sometimes the focus of those opposed to dam construction. Much of the criticism has been based on environmental, ecological and social costs imposed by dam projects and not on technical aspects.

The sensitivity of the World Bank to environmental issues dates to the early 1970s about the time environmental issues were becoming a part of general public discourse. Since then, the Bank has tried to respond to the growing worldwide environmental awareness. With regard to dams, the Bank has produced a series of guidelines and other procedures to help incorporate safety and environment effects more fully into the design of large dams. The International Commission on Large Dams (ICOLD), an international governmental organization, has produced several technical bulletins and publications dealing with dam safety and environment. ICOLD has several times selected these two topics for discussions during its international congresses and symposiums.

The purpose of the seminar held in April 1989 was to consolidate the experience of the World Bank, together with representation of the consulting industry, and government agencies on the design and management of dam projects in order to improve the safety and environment impact. The seminar also aimed at opening a dialogue with some NGOs.

The discussions during the seminar of the experience in the World Bank and of some countries are presented in this volume. A summary description of the seminar proceedings was published in the August 1989 issue of *Water Power and Dam Construction*. This summary is reproduced in this volume with the kind permission of the editor of the publication.

The first part of the proceedings provide an overview of the two main topics. The second part deals with: (i) Bank experiences on dam safety and the implementation of Bank policy; (ii) the experience of developed and developing countries in the field of dam safety; and (iii) the design for dam safety. The third part deals with environmental aspects with special reference to water quality and public aspects.

Valuable contribution to this debate was made by the president of ICOLD and experts from the United States, United Kingdom, Brazil, Canada, France, India and the World Bank.

It is hoped that this publication will provide some guidance to decision makers on the development of their water resources and help to open dialogue with those concerned with the social and environmental impacts of large dams.

Mr. Rajagopalan supported the initial concepts and objectives of the seminar. Several people contributed to its success including its chairman, G. Le Moigne and S. Barghouti, H. Plusquellec and Mrs. S. Thorpe who were responsible for its excellent organization.

Michael Petit, Director
Agriculture and Rural Development Department

A REVIEW OF THE SEMINAR[1]

A two-day internal seminar organized by the World Bank recently (April 17-18, 1989) analyzed the relationship between dam safety and environment, and associated trends in dam construction.

The Vice-President of the Bank's Policy and Research Sector, **V. Rajagopalan**, stressed, in his opening address, the World Bank's objective to "make correct choices, in a more environmentally challenging world, so that the real benefits of dams can be achieved."

J.A. Veltrop, President of the International Commission on Large Dams and Vice-President of Harza Engineering, noted that stress on the environment had increased significantly during the second half of the 20th century. Improvements in living standards and the rising expectations of a rapidly increasing population had necessitated the construction of additional dams for water supply, flood control, power generation and recreation. Although dam safety had always been fundamental to the dam engineering profession, he said, today's concerns about potential failure were increasingly important because of the concentrations of population in downstream areas. He felt that, while it was generally acknowledged that dams were indispensable, the social understanding of the limited nature of natural resources and the associated fragile environmental balance required more attention, so that undesirable environmental impacts could be mitigated.

Veltrop noted that, although water resources on earth were quite abundant, sources of fresh water were limited (2.67 percent), and rivers represented only a small fraction, which required careful management.

---

1. The review was done by *Water Power and Dam Construction* and was published in their August issue. This article is reproduced with the kind permission of the editor.

He called for conservation measures to be implemented simultaneously with the conception and construction of dams. "Lessons have been learned from the detrimental effects of certain dams", he said, "and ICOLD, USCOLD and the World Bank are utilizing these lessons to alleviate the adverse impact of dams."

He observed that dam safety was also affected by the aging of many structures; many dams in the world were now more than 50 years old.

G. Le Moigne, of the World Bank, the Chairman of the seminar, commented that the Bank's policy was to reduce support for those dam and reservoir projects "without adequate safeguards against harmful environmental consequences, particularly those which unduly compromise health and safety."

Le Moigne said that dams, especially those financed by the World Bank, were sometimes criticized on the grounds that their social and environmental costs outweighed their benefits. However, he noted that, in relation to dam building throughout the world, the World Bank was only involved with a small percentage. According to ICOLD statistics, there were more than 35,000 large dams in existence. Some 11,000 had been constructed in the period 1951-82 (an average of 344 dams per year). In contrast, the Bank was the major funding agency for 5 to 10 dams per year, or less than 5 percent of the annual average.

Safety

The session on dam safety focused on the experience of the World Bank in this field, and the implementation of its policies. A presentation on this subject was given by the Bank's dam safety specialist, **P.N. Gupta**, who described the World Bank Guidelines on the Safety of Dams. These, he said "provide a solid base for safer dams", although he felt that issuing these guidelines alone would not be adequate to ensure that a project was safe. He summarized essential steps for dam safety as follows:

1. implementation of Bank guidelines;

2. involvement of Bank staff (with knowledge of dam safety aspects) at various stages of feasibility/design/construction;

3. procurement and contract packaging to suit safe implementation of works;

4. co-financing packaging aimed at safe implementation of works;

5. realistic, effective and successful implementation programme including dam instrumentation;

6. timely resolution of technical issues;

7. project to be treated as a dam project, with consideration of dam safety aspects;

8. continuity of professionals on the project - in the Bank, the country of implementation, and the project consultants;

9. participation at relevant meetings;

10. training of local professionals for dam safety aspects, instrumentation and monitoring.

Y.K. Murthy then discussed Indian experience in the field of dam safety, and B. Price outlined specific actions to be taken during the planning and construction processes. Various consulting engineers commented on the safety aspects of dam projects in different parts of the world.

Asked about making a distinction between developed and developing country policies on dam safety, **L.A. Duscha**, expert on safety of the US National Committee on Large Dams said: "Although developed nations do take responsibility for dam safety, none has achieved perfection." He added that it was more significant to note the generally recognized differences in the treatment of dam safety between the American federal and non-federal sectors than between developed and developing societies.

Duscha said that dam designers over the years had been concerned with safety, but that safety encompassed more than simply design; it also had to be associated with planning, construction, operation and maintenance. He noted that, although much neglected, programming, budgeting, administrative and management facets were also related. "In fact, it is during these phases that the all-important commitment to dam safety must be made", he added.

Although it has been proposed that dam safety be approached from a cost-benefit standpoint, Duscha said, risk analysis to determine the probability of failure required the quantification of several factors. Among these, human error had been judged as the most likely to cause failure, and unfortunately, this factor was not quantifiable. In addition, loss of life could not be expressed in monetary terms. Because the value of human life was a philosophical matter than a direct function of cost, Duscha felt that a cost-benefit approach to safety was inappropriate.

Duscha mentioned that the ICOLD Committee on Dam Safety had conducted a survey on dam safety legislation among the 75 ICOLD member countries. Of the 36 countries which had responded, 20 had indicated that they had no dam safety legislation. Most of these were developed countries, demonstrating that safety had a long way to go to be effectively formalized on a world-wide scale.

W.T. Smith, Chief of the Bank's Energy Division, Asia Region, Said that the tremendous advances in living standards and quality of life throughout the world in the 20th century had been driven primarily by large-scale water resources development, although there had been problems associated with these, such as waterlogging, salinity and the spread of water-borne diseases. However, he pointed out that the areas affected by such problems were small, compared with the vast areas which had remained free of such problems.

Smith observed that the world's present irrigated area of 270×10^6 ha covered nearly 20 percent of the total cultivated land, and provided one-third of its food production. Irrigation from any source (diversion dams, pumping plants, or reservoirs) he said, could cause waterlogging and salinity. Storage dams could sometimes create additional water which was used to increase the intensity of irrigation in established areas. Thus, if the water table rose, the dam could be said to have caused waterlogging. However, he stressed that in most river basins, the amount of stored water used was small compared with total water diversions for irrigation, and reservoirs were seldom a major factor in waterlogging or salinity.

J.J. Ellam, Chief of the Division of Dam safety of the Pennsylvania Department of Environmental Resources, USA, said that the 1970s could be characterized as an "era of awareness in dam safety", when several major failures had increased awareness of the potential hazard that dams created. The 1980's had been expected to be the period for action in the mitigation of the public threat posed by unsafe dams. He said that there had been high and low points in the decade, and dam safety had enjoyed varying degrees of attention. "One would say that the public's attention varies in inverse proportion to the time since, and distance from, the most recent catastrophic event", he observed. "As most engineers described the phenomenon", he continued, "a dam failure is listed as a low probability/high loss event. As managers of dam safety programmes, we would agree that we have to give a great deal of attention to the technical aspects of dam safety, but have been remiss in our attention to the social and public awareness."

Ellam then described progress which had been made in the USA in recent years, since 3,000 dams had been identified as unsafe in 1982. In his state, for example, as a result of an aggressive approach to the issue, a total of 208 dams classified as unsafe had been reduced to 40.

He concluded that the next critical environment-induced dam safety issue would be the aging of structures, as, by the year 2020, 85 percent of the 80,000 non-federal dams would be more than 50 years old.

Designing Safety into Dams

P.A.A. Back of Sir Alexander Gibb & Partners, UK, said that the safety of dams depended on three main factors:

> **First**, it depended on the design - how well the designer had understood the many different facets involved, such as: the foundation geology; the floods to be routed; the materials to be used in construction; and, the durability of those materials in the given environment.

Second, safety depended on the quality of construction: does the quality of construction match the assumptions of the designer, or has sloppy workmanship invalidated the design assumptions? Have poor materials been used in place of those specified?

Third, safety depended on subsequent operation and maintenance of the dam. There was no such thing as a dam that could be left to its own devices. It always had to remain under supervision, and it always needed to be monitored.

Monitoring Existing Dams

D. Bonazzi of Coyne et Bellier, France, stressed the requirement to re-assess the safety of existing dams, adding that specialists had an increasingly important role to play in the field of monitoring, maintenance and the repair of old structures. He commented that there were very few examples of cases where a reservoir had been emptied so that the land could be returned to its former state; there were obvious practical difficulties in draining an impoundment, therefore an artificial reservoir had to be regarded almost as a permanent part of the landscape and environment, (which it had usually profoundly modified).

"No illness can be treated before it is identified", he said, "and this also applies to existing dams, which are complex structures, sometimes designed, built and operated by distant generations who have left their mark and passed on a more or less mysterious heritage." In this context, he said a solid historical knowledge of construction techniques was most valuable. As an example, he said that practically no masonry dams had been built in recent years, yet specialists could be required to know what to expect of such dams when assessing their performance.

Environment

M. **Yudelman**, of the World Resource Institute, commented that recognition should be given to the fact that the overall environmental impact of a number of small schemes could be greater than that of one well managed large scale project. He also noted that dam-related environmental impacts at least did not affect the global environment, as did, for example, factors contributing to the so-called greenhouse effect, such as carbon dioxide emissions.

F.H. Lyra of Brazil, giving an overview of environmental aspects of dams, drew attention to the significant contribution made by ICOLD to this field since 1972, when the first committee had been set up on dams and the environment. Lyra had chaired this committee before taking up the post of ICOLD's President in 1976. In addition, since this time, several detailed discussions on the subject of environmental aspects had taken place at ICOLD Congress, he said, and last year in San Francisco a whole Congress question (Q60) had been devoted to the subject.

An overview of considerations of environmental effects in World Bank-financed projects was presented by **J.A. Dixon**, Consultant to the Bank. The presentation looked also at economic aspects, and stressed that dams did not contribute to the economic development in a net sense unless their benefits outweighed their negative costs (where costs included environmental damage). Dixon pointed out that failing to account of environmental consequences was only to mislead oneself about the contributions of the dams to economic growth. "Whether environmental costs are accounted for or not, they are still there", he said.

He noted that dams were significant social investments, mostly part of larger overall development projects, in which their costs could have different economic significance. For example, within an irrigation scheme, a dam might account for 20 percent of the investment, in an energy scheme, between 20 and 50 percent, and so on. He said that an economic analysis should take into account direct and indirect benefits as well as environmental costs.

Statistics were presented by **T.W. Mermel**, Consultant to the World Bank, on recently completed dam projects around the world, which updated information available in 1988 on major dams in the world. He noted that 45 schemes were under construction, and 12 had been completed. These schemes represented a significant increase in reservoir capacity (550 km^3 for the 45 under construction, and 210 km^3 for the 12 completed).

T. Vladut, of Retom, Canada, then discussed the phenomenon of reservoir induced seismicity.

Specific environmental issues relating to reservoir management and water quality were discussed in papers by **R. Brooks** of the Tennessee Valley Authority (which was presented by M. Davis) and **J. Olivares** of the World Bank.

Brooks described his organization's management of the schemes on the Tennessee river and its tributaries, most of the reservoirs having been constructed for multipurpose use (primarily navigation, flood control and power production). He outlined some of the negative effects of impoundments, for example, intermittent release patterns of high and low flows to meet peak demands for hydro power, which had affected habitat conditions for aquatic life, and caused changes in water quality and lowering of downstream water temperatures.

Brooks reported that TVA was testing and implementing reoxygenation and flow regulation measures at various projects to avoid or mitigate some of the undesirable effects of impoundments, and was re-evaluating its reservoir operating policy to determine whether changes were feasible which would enhance beneficial effects of impoundments for multipurpose use.

Olivares focussed his presentation on the public health impact of large irrigation projects, particularly in Africa, where many water-related diseases had spread or amplified. He recommended various actions to be taken in the form of safeguards to be built into a project's design to minimize, or at least pre-empt anticipated health risks. Some of the strategies he recommended are shown in the Table.

Preventive Strategies for Water-Related Diseases

Disease Category	Preventive Strategy
Water-borne	Improve water quality. Prevent casual use of unimproved sources.
Water-washed	Improve water quality. Improve hygiene. Improve water accessibility.
Water-based	Decrease water contact. Control snails. Improve water quality
Water-related insect vectors	Improve surface water management. Destroy breeding sites. Decrease human-insect contacts.

Source: Adapted from a paper by R. Feachem, M. McGarry, and D. Mara.

Conclusion

The seminar reflected the continuation of the World Bank's increasing interest and concern with environmental and safety-related matters associated with the construction of large dams. It was generally agreed that broader and more reliable approaches in dam development were required, in line with a policy of continuing growth, and in recognition of social requirements for environmental preservation and enhancement.

PART I

OVERVIEW

Chapter 1

DAM SAFETY AND ENVIRONMENTAL SEMINAR

V. Rajagopalan

Man has built reservoirs to exploit water resources throughout historic time. Some very early examples built some two millennia ago still survive in good condition today as, for example, the reservoirs of Anuradapure in Sri Lanka. In the last 100 years with the rapid growth in demand for water for all purposes there has been a dramatic increase in the number of dams. By 1950, there were about 5,000 dams in the world over 15 meters in height. In the last 49 years, the number has increased about six-fold to some 30,000--of which more than half are in China.

An important stimulus to dam construction has been the tremendous growth in domestic and industrial water supply throughout the world. A 100 years ago piped systematic water supply systems were rare even in big cities but now they are commonplace.

The early years of this century saw the beginning of hydroelectric development in those countries that had the right combinations of topography, hydrology, and power demands. In many developed countries, hydro-power has already been fully exploited but a vast unexploited potential remains in Africa, South America, and in many parts of Asia.

The tremendous advances in standards of living and quality of life throughout the 20th century have largely driven by large scale water resources development.

The World Bank has always been deeply involved in water resource projects for all purposes--domestic and industrial supplies, hydroelectric power generation, and irrigation. Approximately one fifth of the Bank's lending has been devoted to water projects serving one or several of these purposes.

Not surprisingly, the large recent program of dam construction has created environmental changes which are giving rise to important and far reaching issues. It is appropriate that the Bank should take careful account of these issues and ensure that they are properly evaluated in all the projects it supports--either directly through project funding or indirectly through program funding. It is through seminars like this that the Bank seeks to exchange ideas, both in-house and with outside experts. Such an exchange not only helps the Bank staff to make better judgments in project appraisals but also guides the Bank's management in the formulation of its policies and strategies involving dam construction.

The seminar deals with two broad topics that are of evergrowing concern in water resource development. These are **the environmental effects of**

reservoirs and **the safety of dams.** The Bank wishes to ensure that for the dam projects with which it is concerned, all reasonable steps are taken to alleviate adverse environmental effects and to exploit the favorable ones. Above all, the design and construction of the dam--and indeed its subsequent maintenance--should be executed in a manner that will provide the maximum safety.

There is today a growing criticism that the adverse effects of dams are not taken adequately int account. Some of the stronger critics allege that if this had been correctly done, many recent dam projects would have been abandoned. The most significant impact focussed upon by the critics is the problems of displacement of populations from the reservoir areas and their resettlement or compensation as the case may be. The problems of displacement of populations and land appropriation have become particularly severe in recent years owing to the shortage of suitable resettlement areas coupled with the increasing population density of the areas to be appropriated. In the early days of irrigation development, each new scheme was a resettlement project in itself and those displaced from areas designated for canal or reservoir construction found themselves in possession of some 10-20 hectares of fertile irrigated land with a much greater production potential than ever they had before. Many of the dams to be built from now on do not lead to much extension of irrigated areas but more to the intensification of cropping on existing areas as, for example, in Pakistan, India, and China. There is thus little opportunity to provide specially designated resettlement areas. This situation is even more acute for hydroelectric projects which are becoming almost socially unacceptable in the well-populated parts of the world.

Apart from the problem of displacing populations, there are an array of environmental effects that call for detailed examination. These include water quality, ecological change, and human disease. Some of the impacts are adverse and some, such as fisheries, are beneficial. The adverse impacts tend to become more difficult to alleviate as countries or regions become more developed - in domestic water supply the pollution from phosphates and other contaminants promotes the enrichment of reservoirs with all the problems of eutrophication and detrimental effects on potable quality. The conversion of irrigation systems to greater crop intensification adds to the incidence of malaria and bilharzia vectors. It is important to monitor these trends and develop the right technology to deal with them.

Dam safety has always been a matter of paramount importance to the Bank's appraisal missions. There are three important steps to ensure the highest standards of security. Firstly, safety considerations must be carefully embodied int he design of the dam structure.

Secondly, the standards of construction must be effectively supervised to translate the design criteria into the finished product, and thirdly, the dam, once built, must be monitored and maintained and operated in the correct manner. I note that each of these steps is scheduled for discussion in the program of the seminar. These remarks are not intended to imply that dam safety standards have been seriously deficient. On the contrary, the record is good by any other engineering standards.

Surveys of dams built in recent time demonstrate a good safety record. For example, in the USA, of the 2,000 high dams built during 1950-1973 there were seven failures and three of those involved dams that were constructed with little or no engineering design or supervision. Safety standards are constantly being improved with better design techniques and higher standards for all its projects, the World Bank requires the special dam safety review panels to advise and review the design and the implementation of the project throughout the construction period. The Bank also seeks to promote national legislation to allow for the regular inspection of existing dams by qualified experts. As the World proceeds to build more than 500 major dams a year, there will be a need for a continuous dialogue between all interested parties in order to allay fears, improve technology, mitigate the adverse effects and ensure that the benefits to mankind are fully and equitably released.

Chapter 2

WATER, DAMS AND CIVILIZATION

Jan A. Veltrop

Abstract

For 5,000 years or more, civilizations have relied on dependable
water supplies for consumption and irrigation. Now, as before,
dams are indispensable to meet these same human demands, as well
to as protect against floods, generate power, improve navigation,
provide recreation, and promote economic development.

Safety of dams has always been a concern of the engineer. Population
concentrations in vulnerable areas below dams have created anxiety among
people and governments as a direct result of well publicized dam failures.
Today, aging of dams is as much a consideration for dam safety as floods,
adverse geologic conditions, earthquakes, seepage and design or construction
defects.

Significant advances in the design and construction of dams have been
achieved over the past 100 or so years, partly as a result of evaluating dam
failures and near-failures, but mostly due to broadening the scientific base
of geology, hydrology and the properties of natural and man-made materials, as
well as a better understanding of the loadings to which a dam is subjected,
new analytical methods, powerful computers, quality control during
construction, and instrument observations of structural behavior of dams.
Factors of safety have been refined, human errors reduced and design criteria
have found international consensus. Independent reviews of designs are
carried out routinely, and monitoring of dam behavior has become standard
practice.

ICOLD has contributed significantly to these advances by means of the
exchange of experience among professionals of many countries, and through
papers and discussions at its tri-annual congresses. The work of its
technical committees and the publication of bulletins are also directed
towards improving dam safety, advancing the state-of-the art and promoting dam
safety legislation in all countries. A special Committee on Dam Safety has
published detailed guidelines for practical use by designers and operators.

Man has affected his environment ever since he introduced agriculture
and started to exploit natural resources. As a result nature could not and
cannot be preserved in an unaltered state. Increased public concern
necessitates drawing a careful balance between beneficial and detrimental
effects in all of man's activities, including the construction of dams.

The need for and the beneficial effects of projects such as the Hoover dam and the Tennessee Valley projects are clear: flood control, new farming opportunities, power generation, and alleviating poor economic and social conditions. As water supply and related economics become more secure in recent 20th century history, survival and economic concerns are being replaced by environmental and social concerns. Much has been written about the adverse effects of dams and reservoirs, and many unsupported accusations have been made. Among them, the suggestion to stop dam construction altogether is impractical and contrary to people's need for water resource development. Such drastic action also is totally unnecessary.

Lessons have been learned from the detrimental effects of certain dams. ICOLD, USCOLD and the World Bank are utilizing these lessons to alleviate adverse impacts of dams. ICOLD's ongoing work was formalized in 1972 with the formation of a committee on dams and the environment. Its first publication included a matrix to be used for listing and evaluating the impacts of individual dams on specific aspects of the environment. Concern and knowledge of engineers on environmental matters was illustrated with a bulletin on the environmental success of specific projects. Histories of projects in Austria, Finland, Sweden, Mali, and the United States were published in 1988. Variations of environmental effects in severe winter, temperate and tropical/arid zones were studied and published, including solutions and recommendations to counter the undesirable effects of large dams.

The stress on the environment has increased significantly during the second half of the twentieth century. Improvements in living standards and rising expectations of a burgeoning world population necessitate the construction of additional dams of any size for water supply, flood control, energy production, recreation. Modern techniques assure the safety of these designs, and experience with environmental effects enables avoidance of unnecessary damage to the environment through careful planning and management.

Public awareness of the importance of preserving natural conditions requires participation of all interest groups in decision making about water resource development, in particular, about dams.

We must conclude that water is essential, power is needed, flood control is beneficial and that therefore dams are indispensable. Dams are safe, the environment can be protected and social and cultural effects are given careful attention. Each project must be evaluated on its own merits and as part of a broader system.

Historical Significance

Early Dams. Throughout history, society has been the instigator of engineering works. Construction of dams became a necessity when the assurance of dependable sources of water became a societal requirement. Dams have influenced the rise and fall of civilizations since early times, especially those cultures highly dependent upon irrigation. Many of the structures failed as subsequent generations lost the knowledge necessary to maintain or reconstruct the dams. Without water these civilizations faded away.

Evidence exists that "dams" were used at least 5000 years ago in the cradles of civilization in Babylonia, Egypt, Persia, Sri Lanka, India and China. In 2900 BC on the Nile near Kosleish a 15m high masonry structure was built for the water supply of Memphis. A 6-m-high rockfill dam built in 1300 B.C. on the Orontos in Syria is still in use. Assyrians, Babylonians and Persians built dams for water supply and irrigation. Because rainfall was ample and well distributed throughout the year, Europe generally did not feel the need for dams until the advent of the industrial revolution. Here, early dams were limited to creating reservoirs for towns, driving water mills, and replacing water losses in navigation canals. Not until this century, however, have dams in many parts of the world become truly multipurpose, with emphasis on power generation and irrigation water supply.

The significant advances in the engineering of dams, made over the past 100 years or so, have resulted in much larger and higher dams, have increased their safety, and have enabled construction to take place on less favorable foundations. Until recently the design and construction of dams was entirely the prerogative of engineers. This is no longer the case. Society has gradually become more and more concerned about macro-economic cost allocations, protection of the environment and the risk involved with living in the flood plain of a dam. Throughout, the fundamental concern of the engineer to build safe and economical structures has remained unchanged.

Water Availability. Although there is plenty of water on the earth, some 1,385 million cubic kilometers, it is significant that its distribution renders more than 97 percent of it unsuited for most human needs.[2]

Item	Sources of Water (approx.) Volume (cu km)	% of Total
Salt water in oceans	1,347,900,000	97.3
Saline lakes and inland	105,000	.008
Fresh water	37,000,000	2.67
	1,385,000,000	100

2. Encyclopedia Britannica, 15th Edition, 1987 Vol. 20; page 789 - The Hydrosphere.

Sources of Water (approx.)

Polar ice and glaciers	28,200,000	2.04
Groundwater < 800 deep	3,740,000	0.27
800-4,000 m deep	4,710,000	0.34
Lakes	125,000	0.009
Soil moisture	69,000	0.005
Atmospheric vapor	13,500	0.001
Rivers	1,500	0.0001
Total supply	36,859,000	2.665

According to United Nations estimates there will be 6 billion people by the year 2000, with more than 50% living in cities. These 6 billion people can survive only if the world's reserves of fresh water are managed much more carefully than at present. There are significant problems with the availability of water for general use. The most important are:

1. Great imbalances between regions, e.g. 1/6 of the total river flow of the world is in the Amazon.

2. Variations in precipitation and streamflow.

3. Unavailability at times and places where needed.

4. Flood runoff often cannot be used economically.

5. Population have grown and regions have developed where readily usable water supplies were inadequate.

6. Much of the ground water is at great depth or in sparsely populated areas.

Effective Use of Water Resources. Although the purpose of dams is well known, one needs to be frequently reminded about the important role they continue to play in providing the basis for civilization. Dams store water to:

1. Meet demands for human consumption, irrigation and industrial uses.

2. Protect against flooding from rivers or the sea.

3. Increase available water and providing head for generating hydroelectric power.

4. Improve navigation by increasing the depth of water in a river.

5. Provide a lake for recreation and fisheries.

The proper development of a river basin is part of the engineer's planning procedures as illustrated by the guiding principles of the past President of my company:

1. Each project must fit into an overall plan of development which will utilize the water resources of the basin so as to provide the maximum possible return in benefits.

2. No project in this plan must block any future potential projects through partial development schemes, regardless of how attractive such schemes may be initially.

When economically justified, alternatives to the construction of dams for the purposes of water supply, power generation and flood control are considered in the early stages of planning. Conservation measures may be implemented simultaneously, when water uses for irrigation are improved, when power consumption is made more efficient through conservation measures, and when reforestation or grassing of upstream lands reduce flood peaks.

Caution is warranted when implementation of alternatives to dam construction depends on changing human habits. Time is needed, sometimes in terms of a generation or more, when long entrenched cultural patterns of natural resource usage need to be changed.

Dam Safety

Public concerns. The benefits of dams can be seen everywhere; however the vital services they provide can also be accompanied by hazards. Originally dams were built in areas remote from population centers. Today concerns about potential failure are increasingly important due to population concentrations in vulnerable but limited areas downstream of dams. This concern is the direct result of dam failures, such as the Buffalo Creek Coal Waste Embankment, Canyon Lake dam, Baldwin Hills, Toccoa Falls, Lower San Fernando and Teton dams in the United States, and accidents abroad, such as Malpasset arch dam in France, Vaiont reservoir in Italy, and earthquake damage to Koyna dam in India. Thus, the safety of dams has become a public concern rather than the exclusive domain of engineers and government officials.

Causes of dam failures. Historically 1% of dams constructed have failed. On average 10 significant dam failures have occurred somewhere in the world in each decade, in addition to damaging near-failures. These failures have resulted from a variety of causes: unpredictability of extreme floods, uncertainties of geologic setting, seepage through foundations and embankments, design and construction defects, and liquefaction under earthquake conditions.

Of some 103 failures reported by the National Research Council in "Safety of Existing Dams - Evaluation and Improvement" [3], the causes were as follows:

		%
1.	Overtopping	26
2.	Embankment leakage and piping	22
3.	Flow erosion	17
4.	Foundation leakage and piping	17
5.	Sliding	6
6.	Deformation	6
7.	All other	6
	TOTAL	100

Aging of dams. Dam safety is also effected by aging. Deterioration of dam structures may be caused by weathering of foundations and construction materials, leakage and frost effects on concrete, alkali-aggregate reaction within concrete structures, and corrosion of metal parts.

Data collected by ICOLD show that older dams have failed or suffered serious difficulties approaching failure more frequently than dams of recent vintage. Improvements are attributed to better engineering and construction of modern dams, especially since 1940. Records also show that failures and difficulties have been more frequent during first filling of the reservoir and in the early years of operation. This has been largely due to design or construction flaws or latent site defects. In the extended period of gradual aging there is a reduced frequency of failures and problems during mid life. The frequency of accidents, but not failures, then increases during later life. Often failures have occurred only after more than 100 years of satisfactory service.

Many dams in the world are now over 50 years old. As a result of new knowledge of hydrological conditions, it has been found that some dams have inadequate spillway capacity to pass the so-called probable maximum flood. Also, earthquake conditions appear to be more severe in some instances than originally calculated, in other cases dams were not designed for earthquake loading conditions. Older dams and reservoirs need to be reanalyzed to ensure that they can still meet current safety standards. A special ICOLD Committee was formed in 1972 to prepare a survey of the effects of aging of dams. In 1984 it published a report entitled "Deterioration of Dams and Reservoirs" [4].

Progress in dam design. Up to the middle of the 19th century design of dams was entirely empirical. Knowledge of the properties of materials and of structural theory had been accumulated for some 250 years; Galileo, Newton,

---

3. Safety of Existing Dams - Evaluation and Improvement, National Research Council, National Academy Press, 1983.

4. Deterioration of Dams and Reservoirs - Examples and Their Analysis, ICOLD, Paris, December 1983.

Hooke, Bernouilli, Euler, Coulomb and others made significant contributions. In the 1850's, Prof. Rankine at Glasgow University successfully demonstrated how applied science could help the practicing engineer. Civil engineering then became an accepted university study. Many noteworthy contributions have been made since and are continuing today, such as:

1. Evaluation of dam problems and failures.

2. Improved understanding of the physical phenomena affecting the stability of dams, primarily the loading conditions on a dam.

3. Improvements of cement and concrete, and their control by specifications.

4. The science of soil mechanics, and the understanding and measurement of critical soil characteristics.

5. Geological applications to dam engineering have improved enormously, especially the exploration of foundation conditions and the development of techniques dealing with adverse effects on dam stability, such as grouting, drainage, and seepage control.

6. Development of the concepts of probable maximum precipitation (PMP), unit hydrographs, and probable maximum flood (PMF).

7. Development of mathematical tools to express many physical phenomena in numerical form.

8. Introduction of powerful new analytical methods, backed up by development of computers to solve an array of difficult mathematical problems.

9. Development of effective construction techniques, hand in hand with new equipment.

10. Incorporation of results of instrumentation programs.

11. Observation of seismic behavior.

12. Quality control during construction.

13. Inspection by experienced field engineers.

14. Utilization of the judgment of experienced engineers in the use of design criteria.

The criteria to be applied to the design and construction of dams must reflect the expert interpretation of all site specific data and conditions. These criteria have evolved with time, based on records of performance of dams and the development of technologies allowing the introduction of refined and more complex methods of analysis. Insights and experience on safety issues have been drawn both from successfully completed, operating dams and from dam problems and failures. Design criteria for dams continue to evolve to reflect new experience. Those criteria which have been recognized internationally as indispensable, are most likely to assure safety and reliable performance of dams of all types. An example of this is the ICOLD publication "Dam Design Criteria - The Philosophy of Their Selection".[5]

Design codes for dams must be used with great care, for two reasons. One, they must be regularly updated to reflect the state of the art for each type of dam. Secondly, only experienced engineers should apply such codes to ensure proper interpretation of special conditions or even modification of some criteria specified in such codes. Conditions related to hydrology, geology, seismicity, foundation and construction materials, as well as the proposed functions of the project and the requirements of the owner make it necessary that the engineer approach his assignment with flexibility and utilize judgement based on his experience. In a word, dams are not designed by cookbook.

Independent Reviews. After the failure in 1929 of St. Francis Dam the State of California was the first to pass dam safety legislation. The Judgement by the Coroner of Los Angeles County is noteworthy:

"A sound policy of public safety and business and engineering judgement demands that the construction and operation of a great dam should never be left to the sole judgement of one man, no matter how eminent, without checking by independent experts".

The 1982 USCOLD publication-entitled "Dam Safety Practices and Concerns in the USA"[6] contains an article by Frederick L. Hotes, then Irrigation Adviser to the World Bank, in which he states that since 1977 the Bank has required:

1. A review of the concept and design by an independent panel of experts, acceptable to the Bank, during the early stages of design and during final engineering and construction.

2. Periodic inspection of the dam after construction by suitably qualified independent experts.

It is now an increasingly common practice to have the engineering design of a dam and subsequent construction reviewed by a board of consultants, no matter how experienced and competent the design organization is.

In its loan agreements, the World Bank requires the establishment of a dam safety panel: (i) to review the adequacy of plans and designs of dams, spillways, tunnels, powerhouses, and related structures and facilities, and

---

5. Dam Design Criteria - The Philosophy of Their Selection, ICOLD Bulletin 61, Paris 1988.

6. Dam Safety Practices and Concerns in the USA, USCOLD, Denver, April 1982.

(ii) to conduct periodic inspections during and after construction to determine whether there are any deficiencies in the condition of such structures, or in the quality and adequacy or methods of operation of the same, which may endanger their safety.

Assuring Dam Safety. An integral part of dam design is the use of Factors of Safety. These have been applied routinely by dam designers to account for:

1. Unexpected loads

2. Foundation conditions which remain undiscovered despite site investigations

3. Inaccuracies and approximations inherent in mathematical analyses

4. Computational errors

5. Variations in material properties

6. Deviations caused by construction conditions

7. Human errors in design, operation, and maintenance

The ICOLD publication "Dam Design Criteria - The Philosophy of their Selection" lists the following possible human errors:

1. Failure to respect critical requirements of the specifications during construction

2. Failure to recognize foundation conditions at variance with design assumptions

3. Failure to interpret instrument readings correctly

4. Failure to carry out detailed inspections

5. Failure to carry out critical maintenance or corrective measures

6. Failure to operate a facility as prescribed.

No matter how competent the engineer and how sound his design, careful inspection and quality control during construction are essential to assure that the dam is built according to specifications. Equally important is to report deviations from assumed conditions immediately in order that effective remedial actions can be taken.

Even if all goes well, it is necessary that the behavior of the dam be monitored by means of visual observations and appropriate instrumentation throughout its life in order to check on such unexpected developments as

settlements, deflections, deterioration of foundations and construction
materials, and seepage flow. It is important tot compare performance with
assumptions and calculations made in the original design of the facility.
Such monitoring is especially important during first filling of the reservoir,
when structures are progressively exposed to the loadings for which they are
designed.

The newest monitoring systems are automated and the readings are
processed by computer to allow rapid interpretation of data, thereby
increasing the time available for remedial action. Because monitoring systems
have proved reliable, they are now being installed in many older dams to
provide a systematic basis for observation. Monitoring and inspection must be
carried out by personnel who know what to look for and who can recognize any
signs of distress. Safety standards have been raised in many countries, but
no regulation can replace competence and experience.

The Work of ICOLD. The general interest in the advancement of the
engineering of dams led to the founding of the International Commission on
Large Dams at the World Power Conference in Berlin in 1929. ICOLD is the
primary organization in the world devoted exclusively to dams:

"The objectives of the Commission are to encourage improvement in the
planning, design, construction, operation and maintenance of large dams and
associated civil engineering works by bringing together relevant information
and studying related questions including technical, economic, financial,
environmental and social aspects".

It consists of professional engineers from 78 member countries,
dedicated to improving the technology of dam design and construction,
advancing the state-of-the-art, and promoting dam safety legislation in all
countries. The ICOLD definition of large dams includes all dams over 15 m in
height and those between 10 and 15 m with reservoirs over one million cubic
meters, or crest length over 500 meters, or maximum flood discharge over 2000
cms, or unusually difficult foundation problems, or unusual design.

ICOLD is continuously reviewing the safety of dams by means of a broad
range of topics discussed at its week-long triennial congresses attended by
1400-1600 specialists in the planning, design, construction, operation and
maintenance of dams. A list of these topics is shown in Attachment I.
ICOLD's 16 technical committees (See Attachment II) are working on many
aspects of dam design, all of which contribute to the enhancement of dam
safety. Each committee is composed of 12-20 experts from various member
countries. Publication of bulletins and proceedings from tri-annual
congresses serve as an important exchange of state-of-the-art methodologies
and as an effective feed back of experience to the profession.

A large number of the bulletins prepared by ICOLD involve dam safety. A
representative list is shown in Attachment III. The upcoming ICOLD Symposium
in July of this year in copenhagen has as its theme "Analytical Evaluation of
Dam Related Safety Problems". For concrete dams the subject is the state-of-
the-art of analyzing fracture problems. For embankment dams the topic is:

Advanced analytical models for the prediction of behavior, in particular, concerning erosion by overtopping and piping.

In 1982, a Committee on Dam Safety was convened as a coordinating unit to: define a common safety philosophy, provide an integrated approach for all Technical Committees to address safety issues, propose remedial action if omissions or weaknesses in safety requirements are perceived, and develop general guidelines on dam safety. One of the main objectives of this Committee was to compile and formalize safety considerations for the service life of dams - from panning to final deactivation - and to formulate safety criteria and recommendations that can be used throughout the world. The issuance of such recommendations is an effort to transfer technical knowledge to nations with little or no tradition and experience in dam engineering.

The Committee has now issued "Dam Safety-Guidelines" (Bulletin 59)[7] covering the structural, operational and environmental safety of dams. Legal aspects are dealt with as far as necessary to stress their importance for dam safety. The guidelines can readily be put to practical use by dam designers and operators in any part of the world. They may also be used as a basis for development of national dam safety regulations, by shaping and complementing them according to the prevailing natural, operational and institutional conditions and the particular character an individual country wants to give to its dam safety legislation.

Environmental Considerations

Importance of Dams and Their Environmental Effects. The very existence of humans, animals, and plants has always depended on the availability of water. Man's resources, and as a consequence, nature could not and cannot be preserved in an unaltered state. With the enormous increasing population, dams are even more necessary in today's world then in ancient times. However, it is essential that man conserve nature and its resources and protect the physical environment form avoidable harm and disturbances. On today's crowded planet, it is necessary when considering construction of a dam that full consideration be given towards reaching a careful balance between beneficial and adverse effects. In case construction of a dam is to proceed, these adverse effects must be mitigated to the maximum extent possible.

Water resource projects have many positive environmental effects. When water management practices regulate and augment low flows of rivers and streams, decrease erosion, prevent floods, eliminate waste of water, and in many instances change deserts into gardens where man can comfortably live and prosper, the result is improvement of environmental conditions. Two well-known examples in the United States are the Hoover Dam and the Tennessee Valley projects.

---

7. Dam Safety-Guidelines (ICOLD Bulletin 59, 1987).

Completion of the Hoover Dam in 1935 marked an epoch in reclamation and had enormous technical, economic and social implications[8]. In 1905, a disastrous flood had inundated the Imperial Valley in California for 16 months and increased the previous 22 square mile Salton sink to the 500 square mile Salton Sea. The prime purpose of the project was to alleviate the serious recurring flooding by bringing the widely fluctuating Colorado River under control and transforming it from a menace to a resource. Technically, it was the greatest engineering enterprise undertaken anywhere in the world: a 726-ft-high concrete dam, nearly twice as high as any existing dam, a reservoir eight times larger than the lower Aswan, hydroelectric machinery to develop a million and a third kilowatts, and the All-American Canal to irrigate a million acres.

The dam was a challenge to the imagination, and of paramount importance in developing the nation's resources. The project addressed the basic economic and social problems of the time with nation-wide implications. It became the first project in the early 1930's to aid in solving the problem of unemployment during the years of depression. The benefits of the project were numerous:

1. alleviating the serious economic situation in the 244,000 square mile lower basin area of the Colorado river

2. turning one of the most arid sections of the United States into a rich farm community

3. providing adequate water supply for doubling the irrigated area, and for municipal and industrial use for 12 million people

4. controlling silt

5. improving navigation

6. generating power adequate to pay for the entire cost of the project

7. enabling construction of smaller dams downstream to facilitate diversion of water to Southern California and Arizona.

The Tennessee Valley projects also demonstrated the ability of man to affect his environment favorably. Until 1937 the Valley was a poor district, whose only resources were the growing of tobacco, cotton, and maize. Floods regularly impoverished the soil and reduced its yield. The TVA schemes led to a renewal of agriculture, the development of transport by barges, the construction of electric power stations, the supply of water to the towns and the creation of lakes for recreational purposes. The past 40 years have seen

---

8. Development of Dam Engineering in the United States - Section 3, Arch Dams - USCOLD, Pergamon Press 1988.

great changes in man's management and use of water in the Tennessee Valley. Certainly the development of the TVA reservoir system has helped change the environment in this valley and the evidence shows it has resulted in benefits which overwhelmingly outweigh the losses.[9]

As water supplies and the economy which they supported became more secure, survival and economic concerns waned, to be replaced by environmental and social concerns. Sensitivity to environmental effects is an important development of the twentieth century.

The writers of the "Social and Environmental Effect of Large Dams"[10] have made serious charges against all those responsible for large dams, including engineers, politicians and financiers. In a number of cases concerning some of the most prominent dams in the world, those responsible have been accused of neglecting to study alternatives to water supply, destruction of tropical rain forests, failure of resettlement programs and the increase of malaria and schistosomiasis.

Problems have arisen because of the difficulty in assessing in advance the level of some environmental impacts. However, the presentations by Goldsmith et al; are completely one-sided, frequently inaccurate, greatly exaggerating problems, and completely negative. The conclusion of the book "to stop all dam building immediately" is not only impractical, it is counter to the best interests of people everywhere, especially in the developing countries. More importantly, the world can't afford such drastic action and, besides, it is totally unnecessary.

The World Bank, ICOLD, and USCOLD. In his address to the United Nations Conference on the Human environment in 1972, Robert McNamara, then President of the World Bank, observed: "What clearly needs to be done is to examine the relationship between two fundamental requirements: the necessity for economic development and the preservation of the environment. Economic growth means manipulating the traditional environment. But, on the pattern of the past, it poses a threat to the environment".

His successor at the World Bank, Barber Conable, observed that the Bank must "balance growth with environmental protection". The World Bank requires that environmental and sociological consequences of projects be carefully studied and evaluated as an integral part of project design and optimization, rather than being treated peripherally in separate environmental impact studies. The Bank declines "support for project without adequate safeguards against harmful environmental consequences, particularly those that unduly

---

9. Consequences on the Environment of the TVA Reservoir Systems, R.A. Elliot. Transactions of the Eleventh Congress on Large Dams, Vol. I, paper R15, page 191, ICOLD, Paris, 1973.

10. The Social and Environmental Effects of Large Dams, E. Goldsmith, E. & N. Hildyard, Wadebridge Ecological Center, - Camelford, Cornwall, U.K., 1984.

compromise health and safety".[11] The Bank will assist borrowers "to find appropriate solutions when proposed projects could significantly harm the environment of a neighboring country". With regard to population resettlement and rehabilitation, the Bank requires that plans ensure that people's lives will be improved or at least equal to the standard of living they enjoyed prior to their displacement. Likewise, the Asian Development Bank requires that hydroelectric projects be financially and economically feasible, and environmentally acceptable.

Lessons have been learned from the detrimental effects of certain dams. Sufficient progress has been made to enable engineers and owners to ameliorate adverse impacts of dams and reservoirs. In fact, the effects of dams and reservoirs on the environment have been recognized by engineering organizations for quite some time. During the 11th Congress on Large Dams in 1973, ICOLD members reported extensively on the "consequences on the Environment of Building Dams". Subsequently, delegates have dealt with the effects on dams and reservoirs of some environmental factors, reservoir and slope stability - technical and environmental effects, and in 1988 in San Francisco with reservoirs and the environment - experience in management and monitoring. These are listed in Attachment IV.

In 1978 USCOLD produced a report entitled: "Environmental Effects of Large Dams",[12] which discussed:

1. Environmental control during dam construction.

2. Natural thermal phenomena of physical, chemical and biological changes.

3. The new environment of a reservoir.

4. Algae growth and eutrophication.

5. Aquatic weeds in man-made lakes.

6. Effects of large dams and reservoirs on wildlife habitats.

7. Gas supersaturation problems in the Columbia River.

8. Erosion downstream of dams.

9. Seismic activity associated with large dams and reservoirs.

10. Reservoir clearing.

---

11. Social and Environmental Impacts of Dams - The World Bank Experience. 16th Congress on Large Dams, Transactions Vol. 1, Paper No. 26, ICOLD, Paris, 1988.

12. Environmental Effects of Large Dams, USCOLD, published by ASCE, New York, 1978.

ICOLD Publications. ICOLD is fully aware of the importance and sensitivity of environmental issues and established one of its technical committees to deal specifically with this subject. Since its inception in 1972, the committee has prepared and published a series of four interrelated bulletins, as listed also in Attachment IV.

In 1980, the committee published Bulletin 35 entitled "Dams and the Environment", in which it was noted that the effects of river development schemes may be both beneficial and detrimental on the environment upstream in the storage reservoir, or downstream, or over the whole region. In order to deal with detrimental effects this Bulletin contains a matrix providing a means of listing and evaluating the impact of individual dams and related construction work on specific aspects of the environment. This matrix will enable designers and decision makers to take steps to control detrimental effects and accentuate beneficial ones. The matrix is shown in Attachment V. It takes the form of a table in which the rows deal with the effects on the economic, social, geophysical, hydrological, climatic and biotic environment, while the columns detail the characteristics of action involved, with distinction s between the use for which the water is destined, the type of actin, the zone concerned, physical corrective action and institutional action. By completing this matrix the engineer can develop an inventory of potential environmental problems which need to be addressed. The matrix however does not replace a thorough ecological survey by specialists. On the contrary, it should lead users to seek expert advice on each specific point in order to evaluate the degree to which impacts are harmful and to find ways of controlling them.

Bulletin 37, entitled "Dam Projects and Environmental Success", was published in 1981. It illustrates the concern and knowledge of dam engineers about environmental matters and can be more readily assimilated by non-technical parties. A significant problem discussed in the bulletin is the difficulty of predicting the level of some environmental impacts, even though their possible importance is apparent. Consequently, some mistakes have been made in the past, because, either there has been an underestimation of detrimental impacts or there has been overestimation (usually by public reaction, perhaps arising from a lack of full understanding). The bulletin enumerates most known aspects of possible environmental impacts, together with solutions which have been successful in practice. It describes procedures commonly adopted by dam engineers throughout the world to ensure the environmental safety of their projects.

Environmental effects differ among climatic zones. Bulletin 50, entitled "Dams and the Environment - Notes on Regional Influences", (1985) describes the experiences of engineers involved with projects in the temperate zone, the tropical, subtropical and arid zone, and the severe winter zone. Environmental matters requiring special care within each particular zone are described together with methods which have been used successfully in combatting undesirable effects. Also covered are environmental impact studies, planning and authorization of works, as well as effects on local population, including resettlement and education.

In the temperate zone, particular attention is given to water developments in estuaries. In tropical, subtropical and arid zones, problems of tropical forests and potential health hazard are considered and also resettlement of population. In severe winter climates, ice formation, frazil ice generated by discharge of supercooled water and development of fisheries are discussed.

This 1985 report suggests many solutions and makes recommendations to counter the undesirable side effects of large dams, including:

1. Prevention and control of health problems.

2. Deforestation instead of flooding or burning forests.

3. Rescue of fauna.

4. Forestation in catchment areas to reduce reservoir silting.

5. Financial compensation.

6. Proper water level and water velocity management.

7. Water management to prevent pollution.

8. Construction of fish ladders.

9. Fish breeding and stocking.

10. Planning of recreational facilities.

11. Creation of nature reserves.

12. Reconstruction of natural environments and reforestation or soil restoration measures.

The five projects described in Bulletin 65, entitled "Dams and the Environment - Case Histories" (1988), consist of dams and reservoirs which have been in service for quite some years and therefore provide a realistic picture of the performance of each scheme as related both to achievement of the main project purpose and to the various environmental impacts involved. These projects are located in the different climatic areas of Austria, Finland, Sweden, Mali and the United States.

In Austria, the large-flow, low-head projects on the Danube and Inn Rivers gradually grew into a cascade of power schemes, necessary both for power generation and river morphology. This resulted in a fundamental change of the whole river basin. Adverse effects of earlier structures were remedied. Riverine lowlands were irrigated to ensure optimum conditions for the humid riparian woodlands.

In Finland, extensive environmental studies were carried out for the Lokka and Porttipahta lakes north of the Arctic Circle, in order to predict the environmental impacts of the lakes. These studies included such subjects as flora and fauna, forestry and agriculture, sociology and water quality. Because bogs make up the majority of lake bottoms, initial fears were the possible adverse effects on the downstream water quality and the rising peat floats from the bottom. Monitoring has shown that the water quality in both lakes has improved continuously after the first few years. Changes in the population and subsistence structure included impeding the traditional reindeer husbandry, agriculture and forestry. However, construction of the lakes was favorable to employment and economic stimulus was remarkable with enormously increased fishing and accompanying development of recreation and tourism.

In Sweden, construction of the Suorva Reservoir north of the Arctic Circle was started in 1920. The project regulates the flow of water in the Lule River so that hydroelectric power stations on the river provide 25% of Sweden's total hydroelectric power production. Electricity prices in Sweden are among the lowest in the world, and made it easier for Swedish export industry to compete on the world markets. Construction of the dam and filling of the reservoir resulted in major changes in the landscape and the loss of a beautiful waterfall. The Laps, who are the only inhabitants of the wilderness area, have been provided with a basis for continuing their traditional way of life, such as a number of measures in support of reindeer husbandry, fishing and the construction of the access road. There has been an increase in both the number of reindeer and in the catch of fish. Tourism has developed. At the same time as the hydroelectric power schemes have affected the environment they have contributed to the development of local commercial life.

In Mali, the Selingue' dam on the Sankarani River is a multi-purpose scheme for production of hydroelectricity, water for agriculture, enhanced navigation and fishing. The population was resettled without any problem in new villages, which were provided with adequate services. There has been a fivefold increase in the number of schools. Fishing is abundant, but the planned increase in irrigated agriculture has been delayed because of lack of investment. The spread of bilharzia has been closely monitored and is being kept under control. Environmental aspects were considered an integral part from the very first phases of its design, and the measures adopted are producing excellent results.

In the United States, the Santee Cooper project in South Carolina, was eminently well conceived to meet the needs of the 1930's: jobs in a depressed economy, low cost power to foster regeneration of the State's agricultural community and support for existing industry and the naval establishment. Associated public health concerns and the stress of relocation and resettlement were recognized as well. Not all the effects of the project were as anticipated. Navigation did not revive, but a major striped bass fishery has made the area a great sportsman's attraction. The fish also helped by removing the mosquitoes and the malaria threat they carried. Unexpected siltation of the Charleston harbor has been solved by a rediversion project. Well planned projects such as this one remaining flexible in response to a changing set of cost and benefit criteria, can be successful over a long life. Unexpected problems can be corrected, and benefits, including unplanned ones,

can be enhanced and enjoyed even more. Thus, Santee Cooper means even more to
its people today than when it was under taken 50 years ago.

Concluding Remarks

Water. The stress on the global environment has increased significantly
as a result of rapid population growth supported by historically unprecedented
increases in agricultural production. Rising expectations and improvements in
living standards have given rise to more pollution and deforestation. The
world is facing a steadily increasing demand for water. Global water use
doubled between 1940 and 1980 and is expected to double again by the year
2000, with two thirds of the projected water used for agriculture. Eighty
countries, with 40% of the world's papulation, now suffer serious water
shortages. Although frequent water shortages are an historical fact, there is
sufficient water for the world's projected population, provided it is properly
developed and managed.

Dams. The course of civilization has been influenced by dams since
early times. Civilizations have risen and fallen with availability of water,
especially for irrigation. The well-being of mankind continues to be closely
linked to water supply and dams. More dams, of whatever size, are needed to
regulate the hydrological cycle and to provide a reliable source of water
supply, especially in the developing countries. Without dams people would be
faced with years of hunger, alternating with years of flooding.

Safety. There is no question that dams designed and built today are
safer than ever before and that more is know about how to keep them safe. The
modern well designed and constructed dam presents a negligible risk to the
public. Steps are being taken to strengthen older dams against possible
failure due to earthquakes. The design of large impounding dams in places
with population centers close by requires the utmost in safety engineering.
It is the job of the engineer, the owner, the financing agency and the
government agencies concerned, to reduce the risk of dam failure as much as is
humanly possible. Most governments now exercise statutory control over
engineers qualified to design and inspect dams.

Environment. The implementation of large water development schemes is a
prerequisite for economic and social progress. At the same time, it is
necessary to preserve and protect wild areas where nature has its won way,
where plants and animals can develop freely, and where mankind can breathe and
escape from the stresses imposed by the modern world. A multiple use of
integrated resource management approach is needed. Public awareness of
detrimental effects on the environment is now substantial. It is essential to
undertake environmental studies in the early planning stages of water projects
and to implement the resulting recommendations during the design, construction
and operational phases. Careful planning and management can prevent
unexpected and unacceptable environmental effects. Prevention also is less
costly than rehabilitation.

Public Involvement. Historically, special interest groups have always participated in the planning of water projects but only as direct beneficiary groups. Today, the number of interest groups has multiplied to the point where the planner must be a mediator and an educator. He must understand social and political processes. Informed consent by numerous public groups is a necessary condition for progress in water resource development today. Projects can not be based simply on technical feasibility, but must be based on the values and perceived needs of constellations of interest groups. This requires that social, economic and cultural impacts and resettlement programs should be studied by professionals in these fields at the same pace as the dam project.

Above all, what is needed is a change in public strategy and perception. Water planning and management should not continue to be a battle to be won by a particular interest group but a common effort to solve problems for the benefit of all.

In conclusion, water is essential, power is needed, flood control is beneficial - therefore dams are indispensable. Today we have the knowledge to construct safe projects which take into account all aspects of environmental, social and cultural impacts. Each potential project must be evaluated on its own merits and as part of a broader system.[13]

---

13. Our Common Future, The World Commission on Environment and Development, Oxford University Press, Oxford, New York, 1987.

Attachment I

Dam Safety Related Topics Discussed
at ICOLD Congresses

1933 -   Deterioration by aging of the concrete of weight dams

1948 -   Uplift and resulting stresses in dams

-    Methods and instruments for measuring stresses and strains in earth and concrete dams

1951 -   Methods for determining the maximum flood discharge that may be expected at a dam for which it should be designed

1955 -   Economics and safety of different types of concrete dams

1958 -   Observations of stresses and deformations in dams and their foundations and abutments; and a comparison of these observations with computations and tests on small scale models

1964 -   Results and interpretation of measurements made on large dams of all types, including earthquake observations

1967 -   The safety of dams from the point of view of the foundations and the safety of reservoir banks
-    The behavior and deterioration of dams

-    Dams in earthquake zones or other unfavorable situations

1970 -   Supervision of dams and reservoirs in operation

1976 -   Leakage investigations and drainage of dams and their foundations

1979 -   Deterioration or failures of dams

-    Seismicity and seismic design of dams

1982 -   Safety of dams in operation

1985 -   Dam and foundation monitoring

-    Rehabilitation of dams to ensure safety

1988 -   Design flood and operational flood control

Attachment II

Committees of the International Commission
on Large Dams

Computational aspects of analysis and design of dams

Seismic aspects of dam design

Hydraulics for dams

Materials for concrete dams

Materials for fill dams

Technology of dam construction

The Environment

Dam Safety

Monitoring of dams and their foundations

Sedimentation of reservoirs

Dam aging

Mine and industrial tailings dams

Design flood

Bibliography and information

World register of dams

Glossary and dictionary

Transnational rivers

On the presidency

Attachment III

ICOLD Bulletins Related to Dam Safety

Bulletin 41 (1982)   -   Automated Observations for Safety Control of Dams

Bulletin 45 (1984)   -   Manual on Tailings Dams and Dumps

Bulletin 46 (1983)   -   Seismicity and Dam Design

Bulletin 47 (1983)   -   Quality Control of Concrete Dams

Bulletin 49 (1987)   -   Operations of Hydraulic Structures of Dams

Bulletin 52 (1986)   -   Earthquake Analysis for Dams

Bulletin 53 (1986)   -   Static Analysis of Embankment Dams

Bulletin 56 (1986)   -   Quality Control for Fill Dams

Bulletin 58 (1987)   -   Spillways for Dams

Bulletin 59 (1987)   -   Dam Safety - Guidelines

Bulletin 60 (1988)   -   Dam Monitoring - General Considerations (replaces bulletins 21 and 23)

Bulletin 61 (1988)   -   Dam Design Criteria - The Philosophy of Their Selection

Bulletin 62 (1988)   -   Inspection of Dams Following Earthquake - Guidelines

Bulletin 63 (1988)   -   New Construction Methods

Other bulletins now under preparation include:

Bulletin 44a   -   Tailings Dams and Bibliography (update)

Bulletin 67   -   Sedimentation

Bulletin 68   -   Monitoring of Dams and Foundations

Under preparation is "The Analysis of Statistics on Dam Failure".

Under consideration is the preparation of recommendations on safety improvements of existing dams.

Attachment IV

Environmental Problems Discussed at ICOLD Congress

1973 - The consequences on the environmental of building dams

1976 - The effects on dams and reservoirs of some environmental factors

1982 - Reservoir and slope stability - technical and environmental effects

1988 - Reservoirs and the environment - Experience in management and monitoring

ICOLD Bulletins Related to Environmental Effects

Bulletin 35 (1980) - Dams and the Environment

Bulletin 37 (1981) - Dam Projects and Environmental Success

Bulletin 50 (1985) - Dams and the Environment - Notes on Regional Influences

Bulletin 65 (1988) - Dams and Environment - Case Histories

Bulletin 66 (1989) - Zuyderzee Damming (in printing)

Under preparation are a series of state-of-the-art papers dealing with social, geophysical, water quality, flora and fauna aspects. Another bulletin on case histories will be prepared also.

Chapter 3

DAMS : HISTORICAL AND GEOGRAPHIC PERSPECTIVES

William T. Smith

Introduction

Dams, especially those financed by the World Bank, are being criticized in some quarters on the grounds that their social and environmental costs outweigh their benefits. A summary of the case against dams is presented in a note prepared by the National Resources Defense Council at the request of the Bank. The authors of the note present a wide range of social, economic, and technical objections to large-scale, multipurpose water resource developments. Dams, in general, are criticized because they are key elements of such developments. It is not clear if the objections are only to storage dams or whether they apply also to diversion dams. But consistency would require that all dams are in question since the note is particularly critical of irrigation schemes, many of which depend on diversion dams. Therefore, this response to the NRDC note deals with the broader topic of water resource development rather than a narrow concern with large storage dams.

Promoters, builders and operators of water resource projects see their mission as creating a more favorable environment for the conduct of human affairs. Those who have spent their careers in development would find it hard to accept that the benefits of their work are outweighed by social and environmental costs, or that alternative sources of water and power have been often overlooked. On the other hand, many in the business of water resource development acknowledge that some projects should never have been implemented, that the planning and execution of others has been deficient, and that concern for the human race has sometimes overshadowed concern for other forms of life. Many disciplines are now involved in deciding whether or not a project should proceed, and in the planning and design of development projects. Unfortunately, the partnership between the engineers and the environmentalists is an uneasy one. Each accuses the other of misconceptions, ideological biases, and often just plain ignorance. There is a need to improve communication between disciplines. Alas, most efforts in this direction take the form of an exchange of case histories; alleged failures are contested by reputed successes. In other words, it is debate by anecdote. This paper is an attempt, with limited recourse to specific cases, to put in perspective some of the main issues that have been raised in the debate on dams and their environmental impact. Given the author's many years on the engineering side of development it cannot claim to be unbiased. But it is hoped it will help the reader to find a way through some of the more controversial issues we face today in the development of water resources.

Historical Perspective

Mankind has exploited rivers since prehistoric times. The remains of the earliest organized societies are found along some of the great rivers of the world such as the Indus, the Euphrates, and the Nile. The past hundred years has seen more river development than in all the preceding years of human history. The modern era of irrigation can be said to date from the early years of the 19th century with the construction of the first major diversion dams in India to irrigate several million ha. These were soon followed by barrages on the Nile to provide better water control for two million ha of irrigated land. By the late 1930's, at the start of World War II, projects in the great river basins of Asia and the Middle East had transformed vast areas of desert into irrigated farmland supporting millions of families. With the return of peace, the world's irrigated area continued to expand, especially in China. Undoubtedly, these developments have contributed to the population problems of the developing world. But would it have been a better world if the rivers had been left undeveloped so that population growth could be held in check by famine? Irrigation has not been confined to the developing world: the USA, the USSR, and Japan are now among the leaders in terms of irrigated area. The earliest storage dams, to even out the seasonal variations in flow and provide a more dependable water supply for irrigation, were built in Egypt and India in the late 19th century, and some major storage dams were built in the USA in the 1930's. By 1950 the world had more than 5,000 dams over 15 meters in height, but since 1950 a further 29,000 dams have been built, of which 18,00 have been built in China.

An important stimulus to dam building has been the tremendous growth in demand for domestic and industrial water throughout the world. A hundred years ago piped water systems were a rarity in even the great cities of the world but now they are commonplace. For many of these systems, river water has to be stored by dams to provide a dependable supply, and most of the urban population of the world now depends on reservoirs for their water.

Without the exploitation of rivers the world would be a very different place. Life for many of the people in the great river basins of the world would be cycle of drought, floods and famine. In the days before irrigation, a severe drought in China or India could kill a million people. Without dams and reservoirs, the towns and cities of the world would face desperate water shortages. True, some cities rely heavily on groundwater, but aquifer depletion is now widespread and stored water is needed increasingly to compensate for overpumping of groundwater. The electrical energy produced by the world's hydroelectric projects has supplanted the combustion of vast quantities of oil and coal. If the prediction of a greenhouse effect is true then hydroelectric power has certainly delayed its onset and bought some time to reverse the effect.

The economies of developing countries such as China, India, Indonesia, and Pakistan are heavily dependent on large scale water resource developments. Irrigation provides most of their food and fiber, and their cities, towns and industries could not survive without reservoir supplied water. Some of the fastest-growing economies in the developing world such as Korea, Thailand, and Malaysia are based on the development of their rivers.

In summary, the tremendous advances in standards of living and quality of life throughout the world in the 20th century have been driven primarily by large-scale water resource development, including the construction of large dams. There have, of course, been problems such as waterlogging, salinity, and the spread of waterborne diseases. But the areas affected are small compared to the vast areas which have remained free of such problems. The works of the past cannot however, meet the growing demand for water in the future and, therefore, further water resource exploitation is needed especially in the developing countries. Unfortunately, such development must take place in a more crowded world. Planners now have to deal with river basins where development in the past creates the potential for conflicts between competing projects in the future, and where major works often will face more difficult social and ecological problems than in the past.

Irrigation

It is estimated that the world's present-day irrigated area of 270 million ha covers nearly 20 percent of the world's cultivated land and provides one-third of its food production. Asia has two-thirds of the world's irrigated area. The countries with the largest irrigated areas are:

	Irrigated Area	
	(million ha)	(% of cultivable area)
China	45	45
India	39	25
USA	22	11
USSR	17	7
Pakistan	14	70
Iran	5.8	37
Indonesia	5.3	32
Mexico	5.0	21
Japan	3.2	58
Egypt	3.0	100
Thailand	2.6	15

A breakdown between surface water and groundwater irrigation is difficult to arrive at because most of the best groundwater is in areas commanded by surface water canal systems. In fact, the presence of groundwater is commonly a result of recharge from surface water irrigation. The past 20 years has seen tremendous growth in tubewells. These are nearly all owned and operated by farmers and very few are publicly owned. China, India, and Pakistan have hundreds of thousands of wells.

Waterlogging and Salinity

Waterlogging and saline lands existed in many river basins long before they were irrigated, and severely affected lands were usually excluded in laying out new irrigation systems. Irrigation can, however, cause water tables to rise and create a waterlogged condition. A high water table occurs where natural or man-made drains are inadequate to remove excess rainfall or irrigation water. If the water table is too high, dry-foot crops such as wheat, maize, cotton cannot be grown since they cannot tolerate a saturated root zone. On the hand, a high water table presents no problem for rice production. Land is considered to be waterlogged when the water table is at or very close tot he surface throughout the year. Seasonally high water tables are not uncommon in rice growing areas. A good system of open drains will often cure a waterlogged problem. But in very fine sands the bottom of the drain tends to "boil" and fill the drain and in such cases it may be necessary to resort to subsurface drains such as tile drains. The risk of waterlogging in a newly-irrigated area is hard to predict and the best course is to provide a primary surface drainage system initially and to adopt further measures if a problem develops. Farmers can't be expected to maintain field drains to prevent a problem which may or may not occur for years into the future. In areas with useable groundwater, tubewells provide an excellent solution to waterlogging since the pumped water can supplement surface water. In most countries, the area of waterlogged land has been declining rapidly in recent years because, in areas where there is fresh water aquifer, the water tables have fallen through pumping from tubewells to supplement surface water.

Salinity occurs under high water table conditions. Evaporation removes water at the soil surface and salts are left behind. In its worse form salinity can render land totally unproductive. The basic solution is to install drains to lower the water table and to irrigate in sufficient quantities to leach the salts from the soil. Lowering the water table in saline areas produces a saline effluent which has to be disposed of. In practice, reclamation of severely saline land has proved difficult and costly in saline groundwater areas. However, moderate to light salinity has been dealt with effectively on a large scale in areas underlain by fresh groundwater.

Irrigation from any source--diversion dams, pumping plants or reservoirs--can cause waterlogging and salinity. Storage dams sometimes create additional water that is used to increase the intensity of irrigation in established areas. Thus, if the water table rises the dam can be said to have caused waterlogging. But in most river basins, the amount of stored water used is small compared to total water diversions for irrigation and reservoirs are seldom a major factor in waterlogging or salinity. It should also be recognized that irrigated land permanently lost to waterlogging or salinity is usually land that became productive only as a result of irrigation. Thus, it is not a loss to an economy but an annual gain which could not be sustained. And it should also be noted that waterlogging and salinity were problems in many river basins long before any dams had been built. Finally, it should be noted that vast areas of land in the humid areas of the world, largely devoted to rice production, are free from salinity despite high water tables. And many of these areas have year-round irrigation with the help of large storage dams.

Irrigation and Equity

It has been argued that irrigation benefits the rich at the expense of the poor and it's even been said that multinational corporations are the prime beneficiaries! In fact, most of the irrigated land in Asia (two-thirds of the world's total) is divided into small family-owned farms. This is true even in China where, although ownership rests with the State, many plots are operated on long-term leases by individual farmers. In most countries the problem is fragmentation rather than concentration of ownership.

One of the arguments against irrigation is that farmers are impoverished by being forced to grow cash crops instead of their own food. This is obviously far from the truth because most of the irrigated land in the developing world is devoted to foodgrains. But there are other reasons why this argument is untenable. First, most of the developing world's farmers live in market economies where food is readily available for purchase by those who have income from their cash crops. Second, farmers have considerable freedom of choice in what they grow and the main irrigated cash crops such as cotton, oil seeds, and sugar cane are profitable. Third the world's main cash crops, other than cotton, such as rubber, oil palm, coconut, coffee, tea, and tobacco are not irrigated. Also, it is of interest to note that relaxation of the "grain first" policy was central to the reforms in China which have done so much to raise farm incomes.

The argument has been advanced that "dam-related irrigation projects often lead to mechanized monocultures, which may not be sustainable". This implies that the source of water--whether it be from a dam or a pumping station--somehow determines the type of agriculture practiced! The principal monoculture associated with irrigation is rice cultivation which has certainly turned out to be sustainable but very rarely mechanized. Irrigated land throughout the developing world shows remarkable diversity. Certainly there is a growing trend to mechanization, especially custom land preparation, as farmers become more affluent and value leisure more than they did in the past. Mechanization also leads to more efficient water use and more timely planting and harvesting.

Irrigation Efficiency as an Alternative to Dams

In most of the world's river basins, further expansion of irrigation will require dams to store wet-season flows which at present flow to the sea unused. Those who are against dams argue that they would be unnecessary if water was used more efficiently. The efficiency of an irrigation system can be defined as the crop irrigation requirement divided by the amount actually diverted at the canal head. The difference between the two amounts are the system losses. The sources of potential water loss are: a) seepage losses from canals, b) non-beneficial evapotranspiration which occurs when irrigation water spills into drains or areas not devoted to crops, or when excess water is applied to a cropped area. Water delivered to the crop is consumed by evapotranspiration and by percolation through the root zone: the latter is sometimes counted as a loss but this is incorrect and tends to exaggerate the apparent inefficiency of a system.

Efficiency within a river basin is usually higher than the efficiencies of individual projects within the basin. This is because drainage from upstream projects is used in downstream areas. The rapid spread of cheap, portable, low lift pumps and of private tubewells has led to significant improvements in basin efficiencies. Low-lift pumps are used to lift water from drains and thereby re-use the losses from upstream projects, and tubewells recover seepage losses. Also, studies of irrigation efficiency usually show that efficiency is low in the wet season when supply exceeds demand, but much higher efficiencies are observed when water is short during periods of peak demand. Thus, the scope for efficiency improvements is possibly much less than is generally realized.

In river basins without storage, the river flows are usually concentrated in a few months during the wet season and are uncertain and erratic. Storage of surplus wet-season flows distributes the flow more evenly over a longer period; this extends the growing period in the wet season and, in some cases, permits dry-season cultivation. This function of storage dams would in no way be reduced or eliminated by more efficient wet-season water use.

Hydroelectric Development

Data from the UN Energy Statistics Yearbook for 1986 in Table 4.1 shows the contribution of hydroelectric energy to total electricity production for the world, the continents and the 15 leading power-producing countries. The world's hydroelectric plants produced 20% of the world's electrical energy or 2,000 billion kwh. This is equivalent to nearly 80% of the total production of electricity in the USA in 1986, and is three times the electricity produced in Japan. This renewable and non-polluting form of energy replaces the annual consumption of 240 million ton of coal equivalent. It is inconceivable that in any ecological accounting the yearly production, transportation, and combustion of such vast quantities of fossil fuels would find favor over the alternative of hydroelectric energy.

Although the USA and USSR are, after Canada, the largest producers of hydroelectric energy, their demands are so large they have had to resort to enormous quantities of coal-and-oil-based energy and as result the percentage of hydroelectric energy has declined in the past 30 years. Much of the hydroelectric production in these countries has come from large dams in major river basins.

One of the world's great concentrations of hydroelectric projects is in the Alpine areas of Europe encompassing much of Austria, Switzerland, Northern Italy, and Southern Germany. Some very large dams were built to develop this potential. Grande Dixence dam in Switzerland, built in 1961, is the third highest in the world at 285m. Between them, Austria, Switzerland, and Italy have 17 dams exceeding 150m in height and numerous dams over 100m. Despite this profusion of large dams, the European Alps have escaped the financial, cultural, and ecological disasters predicted by those who oppose construction of large dams in the developing world. Most of the hydroelectric potential of the Alps is now developed: Austria now meets 30% of its demand from thermal plants and Switzerland meets 40% from nuclear energy.

Table 4.1
Hydroelectric Energy

	Total Electricity Production (in billion kilowatt hours)	% Hydro
Africa	325	21
North America	3,200	20
South America	370	76
Asia	1,870	19
Oceania	160	24
Europe	2,530	18
USSR	1,600	13
The World	**9,960**	**2**

The largest power producers:

USA	2,580	11
USSR	1,600	13
Japan	672	13
Canada	620	60
China	444	22
Fed. Rep. Germany	406	1
France	343	18
United Kingdom	298	1
India	202	26
Brazil	201	90
Italy	189	21
Australia	127	12
Mexico	97	28
Norway	96	99
Switzerland	55	60
Austria	44	70

Table 4.2

Per capita consumption of electrical energy

	kwh/capita
Norway	23,800
Sweden	16,900
Canada	16,000
USA	10,900
Fed. Rep. Germany	6,800
USSR	5,600
Japan	5,500
United Kingdom	5,400
Korea	2,390
Brazil	1,500
Thailand	510
China	420
India	260
Pakistan	250
Indonesia	180

While the total production of hydroelectric energy will continue to increase as some large projects now under construction come on line, it will account for a smaller percentage of the more rapidly growing total world energy production. Most developed countries have exploited virtually all of their economically feasible sites. Among the developing countries, Brazil and China have many promising sites still to be developed, and there is considerable undeveloped potential in the basins of the Indus, Ganges, Salween and Mekong. Unfortunately, one of the main problems standing in the way of many hydroelectric developments in the future will be the large populations which have settled in the potential reservoirs.

Energy Conservation

More efficient use of electrical energy is sometimes put forward as an alternative to the construction of hydroelectric projects. Conservation does not, however, eliminate the need for additional power facilities--it simply slows down the rate at which they have to be installed. For example, China has an installed capacity at present of more than 90,000 MW, and an annual growth rate through the year 2000 of nearly 7% ; therefore, additional capacity of some 100,000 MW will be needed between now and the end of the century. With such a low level of per capita use (Table 4.2) the scope for conservation is limited, but assuming that conservation could reduce the growth rate to 5 percent there would still be a need for annual increases in capacity ranging from 5,000 MW to 8,000 MW over the next ten years. Thus, any economically feasible and well-prepared project becomes a candidate for early implementation. This situation is typical of most developing countries seeking to modernize, create employment and raise standards of living.

Alternatives to Hydroelectric Energy

A common criticism of large power--hydro, fossil fuelled or nuclear--is that more benign forms of energy have been ignored. Unfortunately, two of the most commonly cited alternatives--solar, and wind power--cannot, with present technology, be scaled up to be any more than a small component of a modern power system. Wind power also has some severe environmental problems because many of the best sites are scenic headlands overlooking the sea. The location of windmill "farms" at such places would be aesthetically unpleasant and totally destroy their scenic value, and they could also be serious threats to migrating birds. With the latest technology, a 100 foot diameter rotor produces only 300 kw! A 30 MW gas turbine would replace 100 of these machines. It could be placed at the load center and would be environmentally much more acceptable and produce much cheaper energy. Because of the large areas of land needed for solar power plants the best locations are unpopulated deserts in high sunshine areas. These are also often wilderness areas better left alone than disrupted by solar plants and their associated transmission lines.

Small scale, low-head hydroelectric schemes tend to be favored by the opponents of large dams. However, a proliferation of small projects can be much more damaging and disruptive to the environment than a single large project but without the offsetting benefit of a significant contribution to a power system. If countries such as Norway, Austria, and Canada had chosen to develop their rivers with cascades of low-head hydroelectric plants they would long ago have been dependent on fossil-fuelled power and have eliminated vast stretches of natural river channels. Small, low-head projects are not a viable alternative to large hydroelectric schemes.

Gas turbines or combined-cycle plants can be a competitive alternative to a small-to-medium-sized hydropower project where natural gas is available at a reasonable price. They can often be built at a lower capital cost than hydroelectric projects, and this offsets their annual fuel costs.

Dams and Their Impact

The remains of dams on the Nile, in India, China, and in the Tigris-Euphrates basin date back to the beginnings of recorded history. The modern era of water control structures can be said to begin in the mid 19th century with the construction of the first diversion dams in India and Egypt. Until that time most of the world's irrigation schemes depended on inundation canals into which the water simply overflowed from the river; diversion dams or barrages, made it possible to divert larger volumes of water at all stages of river flow. The first storage dams to provide a further degree of water control were the Periyar Dam (1895) in India and the Aswan Dam (1902) in Egypt. Since those early days nearly 35,000 dams higher that 15 meters have been built. Table 5.1. shows the distribution of these by size for the world and for the four countries with the largest number of dams.

Table 5.1

Dams in the World

Height (m)	World	China	USA	Japan	India
15- 30	27,813	16,428	4,212	1,466	881
30- 60	5,537	2,068	827	469	160
60-100	1,095	87	245	163	34
100-150	276	11	44	38	8
150-200	53	1	13	6	2
> 200	24	-	4	-	-
Totals	34,798	18,595	5,338	2,142	1,085

Impact as a Function of Size

The opponents of dams sometimes take the view that high dams present a greater threat to the environment than smaller dams. There are not, in fact any relationship between size and their environmental impact whether favorable or adverse. For example, the barrages of the Indus Basin, none of them higher than 15m have had a profound effect on land and water regimes in the Basin. While most observers familiar with the Basin would argue that their irrigation benefits far outweigh any adverse affects, these dams could be charged with many of the harmful results often attributed to high dams, such as waterlogging and salinity, inundation of land and settlements, and the various social inequities of irrigation. In contrast, the Grand Dixence dam, the third highest in the world, has a negligible impact in terms of the environment.

Physical Effects of Dams

Flow Regimes. The purpose of dams is to divert or store water and, in some cases, to serve both of these functions. Inevitably this affects the natural, unregulated flow regime. For storage dams, the ratio "reservoir capacity to the annual reservoir inflow" is the main determinant of the degree of interference with the natural regime. If the ratio is high, that is the capacity is large in relation to the reservoir inflow, a reservoir can reduce considerably the frequent and magnitude of floods. This can be beneficial if the downstream areas are prone to frequent flood damage. On the other hand, farmers who rely on a flood-dependent crop of rice can be hurt by the absence of seasonal floods. And such hardship and production losses may not be balanced by the other benefits from flood protection. Reduction in flooding can also be harmful to wildlife through damage to, or elimination of, riverain wetlands. There are ways to mitigate such effects but they must be provided for in the design and operation of the project. Many rivers are subject to tidal intrusion in their lower reaches when flows are low in the dry season.

Upstream storage can solve or aggravate this problem depending on how the reservoirs are operated. There are a number of cases where the upstream reservoirs are operated to repel salinity to protect city water supplies and irrigation offtakes. A problem in some basins is that upstream diversions reduce the flow available for salinity control. The solution is then to build a tidal barrage as a physical barrier to saline intrusion.

Sediment regimes. Storage reservoirs retain virtually all of the sediment inflow until the reservoir capacity is so depleted that it trap efficiency is reduced. If a reservoir has a large capacity in relation to the annual flow of the river, it will have a long useful life. But where the capacity/inflow ratio is low and the river carries a high sediment load the reservoir life will be limited. The deposition of sediment in a reservoir can create problems in the river channel downstream of the dam because the sediment-free water released has greater scouring capacity. This results in erosion of the river bed and its banks which can lead to costly protection works and maintenance. It is sometimes argued that the loss of sediment in river flows deprives the land of an important source of fertilizer. However, analysis of most river sediments show them to have a very low nutrient status. A possible benefit is that the sediment can improve the texture of heavy soils.

Water quality. The impounding of water by a reservoir can affect temperature and oxygen content. The temperature of water can be important for irrigated crops especially rice. The problem can usually be solved by multi-level outlets so warmer water can be drawn off when necessary. Outlet works and spillways usually cause turbulent flow and the associated air entrainment often compensates for oxygen lost in storage. Problems of water quality have arisen in warm humid areas when trees and heavy vegetation. Left in the reservoir area have decomposed to produce hydrogen sulphide. This is a considerable nuisance in the vicinity of the dam and can cause corrosion of hydraulic machinery.

Health Effects of Dams

Malaria. The fringes of a reservoir formed by a large dam are, in general, less likely to provide breeding places for mosquitoes than the river channels submerged by the reservoir. The sides of a reservoir are usually too steep to hold shallow water bodies as the reservoir is drawn down. Low dams forming wide, shallow reservoirs can, however, increase the habitat for mosquitoes. Nevertheless, in all areas where malaria is endemic, the fringes of reservoirs should be surveyed to identify possible breeding area and steps taken to ensure that they are drained.

Schistosomiasis. Some of the areas of the world with the highest incidence of schistosomiasis are not irrigated or affected by dams. However, if this disease is endemic in an area it can be spread by irrigation so that storage dams, since they may provide water for perennial irrigation, can worsen the problem. Very few of the thousands of dams that have been built have had this effect. In recent years the disease has been better understood and counter measures have been devised. Thorough surveys and planning of counter measures are needed when dams are built in areas where schistosomiasis is endemic.

Sociological Effects

Land, Property, and People. Most forms of infrastructure--roads, powerplants, dams, irrigation schemes, mines, housing, estates, industrial parks etc.--cause the involuntary movement of people and the acquisition of land and property. Irrigation is one of the few modes of development where new settlements far outnumber the loss of land and homes to dams and canals. However, one of the main problems that governments face in river basin development today is the compensation or resettlement of those displaced by reservoirs.

Fair and prompt compensation is the best way to deal with the reservoir population problem where comparable land is available for those who want to continue as farmers, and where those who want to change occupations have the abilities to relocate themselves and take advantage of new opportunities. But even when these conditions are satisfied there are the problems of land speculation in anticipation of the project, and difficulties in keeping the compensation system free from corrupt and unfair practices. Also, in many reservoir areas, the settlers are recent and do not have the legal titles needed to qualify for compensation.

Resettlement is a viable option if comparable land can be found in large enough blocks to allow whole communities to be relocated under conditions at least equal to their present situation. If properly planned and executed, this approach works if those resettled are ethnically and culturally similar to the people in areas where they will relocate. Problems arise, however, with tribal or aboriginal groups who have close affinity with the environment in the reservoir areas where they live. Resettlement to a different ecological setting can seriously undermine their cultural identity and force them in to an alien life style.

Employment Effects. Large dams have been criticized on the grounds that they cause unemployment because of their capital intensive nature. Presumably what is meant is that dams are often built using construction equipment rather than manual labor. This ignores the fact that dams are built to create employment in industry, agriculture and other forms of human activity and that the faster they are built the faster the benefits will accrue. This is now recognized by most governments, including those that for many years favored labor-intensive construction. It also needs to be said that the construction of a dam by modern methods creates a considerable number of unskilled jobs and provides many people with training in skilled construction trades.

Distribution of Benefits. Dams are also criticized because they are said to benefit a privileged few along with the multinational corporations. The alleged inequity in the irrigation benefits of dams is dealt with in paragraph 3.5. Power and water supply benefits are also widely distributed in most countries through the rapid expansion of electrical systems serving all sections of society. There seems no basis for the charge that multinationals are major beneficiaries of dams. Most are now built by nationally-owned contractors with a large component of locally produced materials. In fact, the multinational involvement in a hospital (drugs, medical equipment) is probably much greater than in a dam.

Effects on Wildlife and Forestry

The inundation of river valleys by dam construction has sometimes destroyed valuable wildlife habitat and rare forestry species. In recent years, much more attention has been paid to surveys of flora and fauna in potential reservoir areas. In some cases rare species have been identified and the projects have been cancelled. In the tropics, most reservoirs submerge some forested areas but the total area taken up by existing and proposed reservoirs is insignificant compared to the destruction of forests through uncontrolled exploitation and land clearing.

There are many cases of habitat for fish and birds being created by reservoirs, especially the shallow reservoirs formed by diversion dams. On the other hand, dams can cut migratory fish off from their spawning grounds. This problem has been handled with some success in North America and Europe through fish ladders and hatcheries. In many countries, irrigation has created excellent habitat for valued species of birds and has helped to compensate for habitat lost to other forms of development. However, much more could be done to make habitat creation an essential part of water resource development projects.

Effects on National Economies

One of the arguments against dams is that their construction costs are so great as to impose an intolerable debt burden on the countries that build them. There may be a few dams where the cost has accounted for a large part of a country's external debt but these are exceptional cases. The two developing countries with the largest number of dams--China and India--are not among the countries considered to have abnormal levels of foreign debt. Furthermore, the foreign exchange borrowed to build a dam could be considerably offset by savings in imported food and fuel, or by foreign earnings from crops and manufactured goods produced as a result of the dam.

Chapter 4

WORLD BANK'S SHARE OF WORLDWIDE DAM CONSTRUCTION

T.W. Mermel

During this meeting we are discussing dams--but no doubt each of you have a picture in your mind of a dam--but are we talking about the same thing? A dam is defined as a barrier built across a water course for impounding water. Some may be high, such as Grand Dixence, Vajont, others may be massive such as Tarbela, Fort Peck, High Aswan, and others irrespective of height create huge reservoirs, such as Kariba, Akosombo, and High Aswan.

We talk about large dams--what is a large dam? By what standard is a large dam large? The International Commission on Large Dams with about 80 member countries is a technical body dedicated to advance dam technology worldwide. It has adopted this definition of a large dam. It must be at least 15 meters high above the lowest foundation, measured to the crest. Under that criteria there are about 35,000 dams worldwide and these are listed in a World Register of Dams. Eighty percent of these are under 30 meters and only 1% are above 100 meters in height.

Since engineers and the public are interested in very large dams, and the additions that take place annually among such dams, a listing of such dams appears in Water Power Magazine, a British publication, under the heading Major Dams of the World. I have been compiling this list for the past 20 years or so and I now update it annually in Water Power Magazine. The criteria for major dams of the world was established to include not only high dams, but also the massive large volume content dams, as well as the huge reservoirs created by dams.

The criteria for this listing is that the dams meet <u>one</u> <u>of</u> <u>the</u> following:

1. The height must be in excess of 150 meters (492 feet); or

2. The volume content of the dam must be in excess of 15 million cubic meters (or 20 million cubic yards); or

3. The reservoir created by the dam must have a capacity in excess of 24 billion cubic meters (or 20 million acre feet).

There are about 350 such dams worldwide.

I also compile each year a Listing of major dams under construction in the current year, and in 1988 there were about 45 major dams underway. Now where does the Bank fit into this picture?

Of the 45 major dams under construction in 1988 the Bank is involved in one way or another, in about 8 such dams. By "one way or another", I mean that the Bank is either funding a portion of the dam construction, or funding the additions to the hydropower installation, transmission lines, irrigation system, or water supply facilities. In no case was I able to identify a major dam being funded 100% by the Bank. As to the 350 major dams of the world, the Bank is or was involved in about 40 dams or about 10%.

Now as to the 35,000 large dams worldwide, in the period 1951-1982 (32 years) 11,000 dams were completed or at an average rate of 340 dams per year. During the 1975 to 1982 period (8 years) the rate of dam construction dropped to about 250 dams being completed per year.

A study made by the Bank in 1985 showed that the Bank was involved, in one way or another, in funding 400 water resource facilities, which were related to a dam, from which the water was used for irrigation, municipal water supply or hydropower generation. We found that in a period of 15 years there were about 26 dams on the average in the pipeline. On further examination, we found that about 8 dams were directly or partially funded, thus excluding those projects associated with dams, which derived water from existing dams for irrigation canals, water supply systems, or hydro-plant equipment or transmission line funding.

When compared with worldwide dam construction of 250 dams being completed each year and with the 26 dams in the pipeline. This represents 10% where dams and related facilities were being funded, and when compared to the eight projects which had dam construction funded by the Bank, this represents about 3 percent of the worldwide dams being completed each year.

Let me cite some specific examples

In Turkey, there are about 219 dams in operation and in 1987, there were 62 dams under construction. Turkey has 39 dams on which final designs are completed, and 34 dams under design study.

Where does the Bank fit in this case?

Of the 62 dams under construction, the Bank is involved in 4 dams.

As to another case I have data on India

India has more than 3000 dams in operation and in 1988 it had 473 dams under construction. Of the 473 dams under construction, the Bank is involved in about 45 dams.

This review is an attempt to distinguish between the many dams which go unnoticed, and which provide the major water supplies worldwide for domestic use and food production. Eighty percent of the worldwide dams are under 30 meters in height. Of the one percent which are above 100 meters high and are funded by their governments, various lending agencies and to a small extent by the Bank.

This analysis is not intended to minimize the Bank's role in worldwide dam construction, but on the contrary the Bank has provided leadership in stressing the importance of dam safety reviews by independent panels of experts followed by periodic safety inspection--also it has provided leadership in showing the way to handle complex resettlement problems involved

in dam construction. Its requirements for environmental reviews as a
prerequisite to dam funding cannot be overlooked. It is hoped that the Bank's
leadership in these areas will also be adopted by the hundreds of other
lending agencies which make the completion of the 250 dams per year possible.

PART II

DAM SAFETY

Chapter 5

DAM SAFETY IN DEVELOPED NATIONS

Lloyd A. Duscha

A basic question arises with the assigned topic-- "Should there be a distinction between developed and developing nations, and if so, why? On the surface it appears that something as altruistic as "dam safety" should command equal attention regardless of the circumstances. Not so, like with most everything else, the realities of the circumstances surrounding the situation enter the equation.

I might add that such differences are generally recognized by the international community of engineers. Also, even in a developed country like the United States there is a difference in treatment of dam safety between the Federal sector and the non-Federal sector. In fact, there is notable difference between the respective states in the United States. And even the Corps of Engineers cannot claim perfection in uniformity.

As I look at it, these differences are driven by a number of factors, which can be categorized as socio-economic-political factors rather that technical factors. Some of these are:

1.   The technical, social and economic sophistication of the country

2.   The size of the country and its economic stature

3.   The tenure of the development stage

4.   The governing economic necessity at the time

5.   The status of enabling legislation concerning dam safety - The power of enforcement or regulation granted responsible agencies

6.   The value of life as derived from the culture of the nation

7.   The cultural approaches to catastrophic events such as Probable Maximum Flood and Maximum Credible Earthquake

8.   The staff capability and the resources dedicated to dam safety

While I have presented a litany of rationalization for differences, this should not be construed as indicating complete tolerance for the situation. Neither should recognizing reality be construed as an excuse for not seeking improvement. The simple statistical facts are that more and more dams are being built and they are deteriorating with time. This in itself could lead to some catastrophic events. Counteracting this rationalization will require continual vigilance on the part of all those responsible for dams and their safety.

Further on statistics, of the 16,000 dams registered by ICOLD through 1975, 108 or 0.7% failed; however, the frequency of failure was significantly lower among recently completed dams. In some countries, 20% of existing dams showed material deterioration. In other countries this was insignificant. Nearly 50% of the failures occurred during construction or first filling.

The probability of sudden dam failure, without previous signs of danger, has been shown to be extremely low. In the 108 failures mentioned above, failure could be traced back to its origin: some failures gave ample warning of danger; in others progressive deterioration could have been detected if monitoring instruments had been used and interpreted correctly.

Regarding the statistical analysis of dam failures, it has been claimed that the number of failures in the next decade can be predicted. This may be so, but who here can predict whose dam it will be. Will it be one of yours? Are you prepared to suffer the consequences?

In the 1970's, several failures and near failures of dams in developed countries strongly indicated the necessity for a more formal and uniform approach to safety. As a result, some nations began an introspection of how dam safety was being implemented. This included reviewing criteria and standards and the legal ramifications surrounding dam safety. Many seminars were held addressing the subject--many papers were written.

Dam designers over the years have been concerned with safety. But safety encompasses more than design. It comes to bear in planning, in design, during construction and carries through the operation and maintenance phase. Although much neglected, the programming, budgeting, administrative and management facets also enter into dam safety. In fact, it is during these phases that the all-important commitment to dam safety must be made. Without a commitment of resources and an appropriate policy, plus a supportive philosophy, the designer, the constructor and the operator cannot discharge their responsibilities for dam safety in a satisfactory manner.

It has been proposed that dam safety be approached from a cost-benefit standpoint. However, a risk analysis to determine the probability of failure requires the quantification of several factors. Among these, human error has been judged as the most likely to cause failure, and, unfortunately, this factor is not quantifiable. In addition, loss of life cannot be expressed in monetary terms. For these reasons, a cost-benefit approach is unsuitable. Since the value of human life is a philosophical matter rather than a direct function of cost, quantification will be influenced by cultural and religious principles, as well as by the level and requirements of the overall national development. Therefore, deliberations about the limits of cost for dam safety may come to dissimilar conclusions in different parts of the world.

Legislation

In an attempt to evaluate the status of dam safety in various countries, the ICOLD Committee on Dam Safety surveyed its member countries regarding dam safety legislation. Of the 75 countries, 36 responded. Of these 36, 20 indicated that they had no legislation. The 16 countries with some form of legislation are: Australia, Austria, Canada, Czechoslovakia, Finland, France, Great Britain, Italy, Japan, Norway, Pakistan, South Africa, Sweden, Switzerland, USA, and Zimbabwe. As the bulk of these are developed countries, you can see that dam safety has a way to go to be effectively formalized on a worldwide scale.

1. **Design Regulations.** Some countries have technical regulations as part of its legislation. Others have codes of practice.

2. **Government-Owned Dams.** While these dams appear to be excluded form the safety legislation in a number countries, some of these jurisdictions have regulations and procedures of their own, which in some cases are more demanding than legislation.

3. **Overview Committee.** In most countries, a government agency or ministry is responsible for implementation of the dam safety program. Some also require a committee to be established to advise on dam safety in general or to address a particular dam safety problem.

4. **Design/Inspection Accomplishment.** Some countries require that design, inspection, operation be performed by properly qualified engineers. By and large, most laws specified intervals between inspections.

In summary, one must conclude that overall legislation is sparse and varies greatly between nations. However, I do not believe one should conclude from this that absence of legislation and non-uniformity are the desired products.

Model Laws and Regulations

As so few countries have suitable legislation to enforce dam safety, one might ask what provisions should be incorporated into a model law to ensure an effective program? I will attempt to outline a few.

Scope. The law should define which jurisdictions have authority over dam safety. In making this decision, it should parallel that of the jurisdiction having the responsibility for the health and welfare of its citizens and the necessary statutory authority--police powers--to enforce and condemn.

Liability. The law must make clear the ultimate liability for damages and the ability to collect for repairs ordered. In nearly all cases, this liability devolves to the owner--whether it be private or public.

Powers. The powers of the agency responsible for dam safety should be clearly defined.

1.   Power to review and approve design and construction.

2.   Features requiring specific approval (plans, specifications, design computations, emergency plans, inspections.)

3.   How and by whom inspections are performed? What should be inspected? Frequencies of inspection.
4.   Data and records to be maintained.

5.   Requirements for professional certification.

Fees. A fee schedule for processing permits to construct or repair, and for necessary government inspections, should be established. This should be sufficient to cover the cost of maintaining a viable dam safety program. The owner of the dam, and not the general public, should be expected to stand the cost, but in certain cases, the Government may have to contribute.

Repair or Removal. Because at times legal ownership of a dam becomes foggy, or the owner does not have fiduciary capability, the government must have the authority to repair or remove a structure. This area poses some complex legal issues and occurs more frequently than one would anticipate.

Complaints. How are complaints of unsafe conditions handled? How are costs adjudicated?

Inspection. The manner of handling inspections of the work during construction and inspection of the completed project should be addressed. Will the government entity perform the inspections with its own staff or will inspections be performed by qualified firms? Will the government entity perform these inspections at own expense or charge the owner? to avoid complications with owners, it is generally preferred to have the government fund the inspection so it can exercise the necessary freedom of decision. A formalized continual inspection program is the heart of dam safety.

Emergency Provision. The agency should have the ability to determine whether an emergency exists and then warn the public. It should also have the ability to take charge and control water releases and take other steps which may be essential to protecting life and property. This should include contingency plans which incorporate inundation maps and evacuation procedures. Some have questioned the propriety of publishing inundation maps and evacuation procedures for fear the public will become unduly aroused. On the contrary, it has been found that the public maturely accepts such plans and appreciates the forthright disclosure.

Penalties. The statute should clearly specify the penalties for various violations and should clearly specify that the responsibility extends to the owners agents and to its employees as well as to the corporate organization. Without the law defining pecuniary and criminal penalties, enforcement will become a mockery.

Standards

In addition to legislation which generally outlines the commitment to dam safety, there is the need for each country, or appropriate subdivision thereof, to provide standards and criteria, which are expected to be met for each dam constructed in that country.

To assist in this effort, ICOLD formed a Committee on Dam Safety. One of the main objectives of that Committee was to compile and formalize safety considerations for the entire service life of dams--from planning to final deactivation--and to formulate safety criteria and recommendations that can be used throughout the world. The issuance of such recommendations as Bulletin 59, "Dam Safety Guidelines," is an effort to transfer technical knowledge to nations with little or no tradition and experience in dam engineering.

The Guidelines cover the structural, operational and environmental safety of dams. Legal aspects are dealt with as far as necessary to stress their importance for dam safety. The guidelines are not viewed as rules or regulations, but rather as recommendations. They can readily be put to practical use by dam designers and operators in any part of the world. However, they may also be used as a basis for development of national dam safety regulations, by shaping and complementing them according to the prevailing institutional conditions and the particular character an individual country wants to give its dam safety legislation.

I have heard it said in both developed and developing countries that there should be separate and distinct standards for existing dams as opposed to new construction. One nears similar views concerning the degree of protection to be provided to withstand rare hydrologic and seismic events. Such statements are usually driven by the high cost of rehabilitation to present standards for existing dams or the added cost in the case of new dams. My attitude is that this is the wrong approach, especially in a developed country. I believe that standards should remain intact so that a common baseline exists. Recognizing that funding is a valid concern does not mean that the standards have to be compromised. I would prefer such decisions be made on the basis of relative risk and affordability. As we well know, there is a risk analysis of some form in most decisions we make. In some cases it is formal, others informal. In some its judgmental rather than mathematical. The important thing is to go through some process to evaluate a decision.

Risk Analysis

Much has been written and spoken about the use of probabilistic risk analysis as a means of making decisions concerning safety standards, particularly regarding the spillway design flood. Some offer it as a panacea to the problem. While I certainly acknowledge its value in assisting to set investment priorities, I find it difficult to become over-reliant on its use. While a probability analysis applied to the risk of rare events may not show an economic advantage for providing the maximum degree of safety, there must be concern for catastrophic failure, even though absolute safety is impossible to achieve. There must be moral concern for the potential loss of life as well as concern for the economic disruption caused by failure. Also involved is the integrity and reputation of the individual engineer and the organization and/or owner responsible for the dam. Can such risks be afforded? I submit very few engineers want to take such risks.

There are numerous parameters involved in determining the relative risk of failure comparing one dam to another. Some of these are:

1. The downstream population which could be impacted by failure.

2. Downstream distance to the centers of population.

3. Percentage of PMF passed by the spillway. (In developed countries, occurrence of this event cannot be categorized as an Act of God. Since it is a recognized possibility, the courts will assume the engineer should have acted accordingly.)

4. Duration and depth of overtopping.

5. Capability of dam to resist the effects of overtopping.

6. Any evidence of structural distress within the dam structure.

7. The potential seismic activity that could be experienced by the dam.

8. The height of the dam.

9. The storage capacity of the reservoir.

10. The impact of the failure on the downstream and surrounding populace and their reaction to failure.

Environment

While on the subject of risk, we should be aware that besides structural safety, the safety of the environment must be taken into consideration. Awareness of the importance of environmental conservation is spreading quickly among developed countries and even to developing countries. Often problems that have become a serious environmental threat have been traced to pure negligence or insufficient funding. In many other cases, environmental problems caused by dam construction have been solved or even developed into secondary benefits.

Complete preservation of the environment is as unachievable as absolute safety. The extent to which environmental problems can be solved is mainly a question of how much society will pay for the benefits of building a dam.

Human Factor

There is a tendency to view dam safety as a technical issue mainly. I submit there is a human element which enters the equation and this element may be the most difficult to cope with. Human error defies analytical modeling as does lack of practical experience in design and in construction inspection, as does lack of technical and administrative competence in the organization that is responsible for dam safety, as does negligent and insufficient maintenance, and as does an inadequate procedure for operation. Any one of these could be a cause for failure. How do we minimize the potential for these human factors manifesting into failures? About the best answer I can provide is that everyone involved has to be continually vigilant during all phases involved--design, construction, operations--to include those crucial interfaces between phases. However, despite one's best efforts, the engineer is always faced with the risk where hidden human error during design and/or construction does not manifest itself for many years.

And, as humans, we should not be deluded that the problem will go away. This problem and its complexities must be effectively communicated to politicians and administrators.

Dam Safety Program

In summary, let me outline what I believe constitutes a good safety program for a developed nation. Frankly, I do not conceive that the goals for developing nations should be much different. Such program should consist of the following components.

1. Adequate and enforceable legislation, including commensurate authority and responsibility, plus the will to enforce the requirements.

2. Recognition that some government entity is responsible for the safety and welfare of its citizens; this entity must provide the inspiration.

3. Owners must recognize their responsibilities and be accountable for the necessary actions.

4. A commitment by the political leadership to provide necessary resources--manpower and money--to sustain a safety program.

5. Some individual has to be in charge of , and responsible for, the dam safety program of the government entity.

6. Incorporate the latest design standards and knowledge.

7. Assure that all designs are reviewed by an independent element, to include the use of outside specialists when necessary.

8. Assure that all projects are operated in accordance with the regulation plan, and are impeded by outside inimical interests.

9. Provide for inspection of construction by a competent well-trained staff.

10. Provide for the periodic inspection of completed projects by experienced teams.

11. Provide for a comprehensive maintenance and rehabilitation program.

12. Maintain necessary expertise through training.

Closing

Although developed nations do take their responsibility for dam safety seriously, none have achieved perfection. The problem will not go away. Our dams will be with us for a long time. Continued vigilance will surely be needed. But probably a greater need will be the monetary investment.

Chapter 6

VIEWS, ATTITUDES AND CONCERNS ABOUT THE SAFETY OF DAMS:
A NATIONAL OVERVIEW WITH A STATE AGENCY PERSPECTIVE

Joseph J. Ellam

Introduction

We are all aware of the benefits of effective dam safety management - benefits as essential as a community's water supply, an area's irrigation, hydropower for near and distant no argument that dams are an integral part of the country's infrastructure. The decade of the 70's may be characterized as the "Era of awareness in dam safety". Several major dam failures increased public awareness of the potential hazards that dams create. The 80's were expected to be the period for action in the mitigation of the public threat posed by unsafe dams.

We have experienced the highs and the lows. Dam safety has enjoyed varying degrees of attention. One would say that the public's attention varies in inverse proportion to the time since and distance from, the most recent catastrophic event. As most engineers describe the phenomenon, a dam failure is listed as a low probability - high loss event. As managers of dam safety programs, we would also agree that we have give a great deal of attention to the technical aspects of dam safety but have been remiss in our attention to the social and public awareness areas.

Tremendous gains have been made in dam safety efforts in the United States during the period of 1970 - 1988. The National Dam Inspection Program of the U.S. Army Corps of Engineers has been the most significant development. the inventory compilation and the inspection of some 8800 dams that had a potential for loss of life and substantial economic damages were milestones. They were considered then as the beginning for what we all hoped would be a continuing effort to identify and mitigate the potential hazards caused by dams.

Yes, we have made significant strides. However, as noted earlier, it is just the beginning. The issues and concerns are many and varied. To quote the Corps of Engineers 1982 Final Report to Congress on the results of the Inspection Program, "in most States, the recommended measures to remedy designated unsafe dams have not been adequately implemented by dam owners, because state officials are not convinced that existing dams create a great enough hazard to require dam owners to spend large sums of money to improve conditions which, in most cases, have existed for many years."

The following are some of the outstanding issues and concerns confronting State dam safety program managers in 1989 as they reflect on the past and look to the future:

1. Adequate levels of funding and support for effective regulatory program management.

2. Availability of funds for dam owners to finance rehabilitation of inadequate or unsafe dams.

3. Consistency in the establishment of design criteria.

4. Public awareness of impact of dam safety programs.

5. The role of risk based analysis in the dam safety decision making process.

6. Inventory data base update and continued maintenance.

7. Liability of state agency personnel in the performance of their duties.

Issues and Concerns

Regulatory Program Management

The National Dam Inspection Program achieved the objective of identifying those dams which posed an immediate threat to human life and property and also fulfilled the intent of P.L. 92-367 to define the national dam safety program.

The unsafe condition of one of every three non-Federal dams and the lack of implementation of recommended remedial measures by dam owners indicate that, for a variety of reasons, most owners are not willing to modify or maintain their dams, and more importantly, most states are not willing to require them to do so.

The Inspection Program was also intended to involve the states in implementing a strategy for developing effective state dam safety programs. The 1982 COE Report states, "The results of this added objective were less than desired."

Surveys conducted in 1982 and 1984 by Bruce A. Tschantz, Professor of Civil Engineering at the University of Tennessee pointed out some interesting facts. The 1982 survey supported the COE findings where it was noted that 21 states had 95 percent of the Nation's total dam safety program budget. Also, only 14 States in the 1984 survey had dam safety programs that incorporated all of the elements of the USCOLD Model Law.

While we are pleased with the improvements accomplished in several of the state's regulatory programs, we are still concerned when there remain seven (7) states that continue to lack the statutory authority for the basic elements of a permitting or regulatory program. Over the last six (6) years, Alabama, and Hawaii had dam safety legislation introduced in their respective Legislatures. Unfortunately, in Alabama, the measure was defeated; however, the legislation was approved in Hawaii. The legislation being considered in each instance was reviewed by the Association of State Dam Safety Officials (ASDSO) and significant input and endorsement was made by the Association.

Also, we have noted instances where a few States with adequate legislation and regulations have been required to drastically reduce their staffing due to budgetary constraints. This is a continuing problem and a matter of grave concern when one considers all of the competing interests for the state budget funds. The "time since the last disaster theory" is very real in the budget process at the state level.

A review of state programs in 1984 disclosed that over half (26) of the States still do not have adequate dam safety legislation and adequate resources and personnel to conduct effective and sustainable dam safety programs. Informal surveys by the writer conclude that this statistic remains viable.

Rehabilitation of Unsafe Dams

The COE 1982 Report indicated that there were nearly 3000 unsafe dams in the Nation. Cost estimates for rehabilitation to bring these structures into compliance with the current state of the art for dam design and construction quickly escalated into the hundred of millions of dollars.

We have seen over the last eight (8) years, a lukewarm attitude towards the implementation of recommended remedial measures for the identified deficiencies. In many states, there is an unwillingness to accept the COE guidelines that resulted in the unsafe dam classification; i.e., the spillway evaluation flood criteria is considered too stringent for existing dams. This viewpoint is shared by State dam safety program managers and owners alike.

Efforts on the Federal level to provide funding assistance have not been successful. There has been legislation introduced and there certainly is interest in the need for infrastructure repair. Despite the rather dim prospects for a major dam repair funding program in Congress, several States have developed and implemented funding programs.

In Pennsylvania, the voters on two occasions have approved state-wide referendums to provide funding for the rehabilitation of water facility related projects. The 1982 program provided loans at rates of six (6) to nine (9) percent and was mildly successful. The present Administration was successful in securing legislative approval for a broad financing initiative that provide funds for low interest loans and limited grant assistance to local communities for financing sewerage and drinking water facilities. The interest rate on these loans will range from one (1) percent to approximately six (6) percent. The program is administered by the Pennsylvania Infrastructure Investment Authority (PENNVEST). The 1982 loan program funded 15 major dam rehabilitation projects with total project costs of $25 million. The PENNVEST program, although it just got underway, has already provided funding for five (5) major dam rehabilitation projects. The amount of loans exceed $23 million with total project costs of $30 million for these five (5) projects.

To say that the above funding provision has been a major factor in Pennsylvania's effort to eliminate the threats to public health and safety posed by unsafe dams would be an understatement. Several states have recently enacted legislation and are implementing loan programs similar to the Pennsylvania program.

So the threat of unsafe existing dams continues unless and until a massive funding program becomes available. How many of the approximately 3000 dams classified as unsafe in 1981 have not been improved. Calculated guestimates seem to indicate that 50%, or approximately 1500 dams, have not been improved.

Design Criteria Consistency

This issue has been one of great interest in the dam safety community. The pros and cons on the continued use of the probable maximum flood (PMF) as the primary basis for the evaluation of spillway adequacy are well documented.

That there is lack of uniformity in dam classification and safety criteria as used by Federal and State Agencies is unquestioned. A review would indicate a wide-ranging variety. As State regulators, we consider these factors as necessary evils but the ambiguous definitions within the current practice (few, more than a few, extensive, etc.) reflect a major issue that needs to be resolved soon. It is noted that it is our responsibility to achieve the proper balance between costs for dam safety repair and the public risks and yet, we cannot develop a uniform approach to the problem. Public policy attitudes, communications among agencies and the variations in engineering judgments are cited as major obstacles.

The recently published National Research Council Report, Safety of Dams - Flood and Earthquake Criteria was expected by many State agencies to provide a definitive statement of appropriate criteria which the Committee discerned as reasonable. This was followed by a comprehensive Report developed by the Surface Water Hydrology Committee of ASCE. The Report, "Evaluation Procedures for Hydrologic Safety of Dams", attempts to identify an acceptable procedure for selecting the spillway design flood which, in addition to the traditional economic analysis, would permit the nondollar-denominated impacts such as loss of life, owner liability, and magnitude of destruction. This Report which resulted form more than three years of intensive effort by a wide cross-section of interests has been widely appraised. Many State regulators continue to be wary since the relative weighing of nonmonetary consequences and their comparison to dollar savings is a matter of judgement which will vary among decision makers for each site examined. Thus, no criteria were provided for making the final safety design decision. In several major states (i.e., major indicates large numbers of high hazard dams), we note that different flood occurrence criteria continue to be utilized for existing dams as opposed to that recommended for new dam designs.

Public Awareness

Responsibility for the regulation of about 95 percent of the dams listed in the inventory rests with the State agencies. We have noted that less than 50 percent of the States adequately address the problem.

Most people knowledgeable with disasters would agree that few man-made structures pose greater potential for destruction than dams. Yet, the public is not convinced that there is a real threat. Within the first few days following a catastrophic dam failure, the initial questions posed by the media are: "Could it happen here?", "Are there any unsafe dams in Pennsylvania/USA?", etc. If the media is uninformed to this extent, public education is and will continue to be a concern.

It is imperative that Federal and State agencies continue with a program designed to improve the public's understanding of the implications of effective dam safety programs. The Federal Emergency Management Agency (FEMA) has provided assistance through ASDSO to initiate and implement such a program.

Risk Analysis

Risk management issues have been a point of concern for several years but only recently has there been a focus on dam safety. FEMA-sponsored research over the last few years has led to a greater understanding by State agency personnel of the use of risk analysis in the decision making process. We note that as long as the potential for loss of life (or cost of life) is omitted from the economic evaluation but included in the decision process, the more likely this approach will be accepted.

We, as State program managers, faced with limited funds have a definite need for this technique and encourage its use and research.

Inventory Update

The 1982 COE report noted that an updated national inventory of dams is a reliable data base for state dam safety programs.

Since 1982, the inventory has been dormant since Federal funding ceased a that point. As a result very few of the States have continued with their own inventory update. Consequently, a large gap in our information base is rapidly developing as new dams continue to be constructed, existing dams are modified, other dams are breached, etc.

It is noted that Federal funding has recently become available and efforts are currently underway for the Corps of Engineers to develop with the 50 States a methodology for developing a consistent National Dam Inventory database.

Liability

I take this opportunity to quote the following from the recent NRC Report - Safety of Dams "most courts strain to invoke liability, particularly when personal injury or death is involved. The odds are substantial that regardless of the theory (negligence or strict liability) cited the result will be a finding of liability in the case of a dam failure involving loss of life."

Surveys of State agencies disclose that many employees are not even aware of the liability risks that their positions place them in on a daily basis. Unless statutory language specifically provides immunity from being sued, there is indeed a large risk. The controversy relating to the appropriateness of spillway design floods for existing dams would seem to be easily resolved by a jury who would consider the PMF requirement of the COE guidelines as the recognized standard of care. If the events of the past are considered to be the prologue to the future, the liability issue will continue to be an important factor in the judgement process.

Views

There are several encouraging developments in dam safety these days. In 1984, several States met in Denver and adopted a Constitution and By-Laws for an Association of State Dam Safety Officials. The stated purpose of ASDSO is to:

1. Provide a forum for the exchange of ideas and experiences on dam safety issues.

2. Foster interstate cooperation.

3. Provide information and assistance to state dam safety programs.

4. Provide representation of state interests before Congress and federal agencies.

5. Improve efficiency and effectiveness of state dam safety programs.

In April 1984, the 12 State Members of the Board of Directors of ASDSO met with Federal officials representing the Interagency Committee on Dam Safety (ICOLDS). This has become an annual meeting providing for a useful and timely exchange of information.

Over the last five (5) years, the Association has greatly expanded and now has 48 States and two territories as paid up members. In addition, Associate and Affiliate Memberships accounted for over 400 additional members. A Newsletter is published quarterly with items of interest. The Association has sponsored National Meetings on an annual basis that generally feature the use of innovative ideas and technology in dam rehabilitation projects. In addition, FEMA has provided funds for initiating several project that have enhanced state dam safety programs. These include the development of a Model State Program for dam safety, implementation of Public Awareness Seminars on a Regional basis, and the establishment of technical groups to consider matters such as the Spillway Design Flood, public employee liability issues, etc.

In my own state of Pennsylvania, we are pleased to report that our dam safety program is making excellent progress. In the National Dam Inspection Program, a total of 749 high and significant hazard dams were inspected in Pennsylvania. The results disclosed a total of 208 unsafe dams (27 percent of those inspected). The primary deficiency cited was the limited spillway capacity of the dams. In part, this may be explained by the fact that about a third of the Pennsylvania dams were constructed prior to 1920.

Because of an aggressive approach and an effective dam safety statute, the Dam Safety program has reduced the number of unsafe dams to 40 as of April 1, 1989. During the 1989-1990 construction seasons, it is anticipated that nearly $40 million will be expended on dam safety related projects bringing the total sum expended in Pennsylvania on public and privately-owned dams to the $100 million level.

It is also noted that several other States report similar experiences. The development of Owner's Manuals for effective Operation and Maintenance of Dams has been or is being implemented in more than a dozen States. Individual States are sponsoring one and two day workshops related to dam safety issues and training. These meetings are well attended by Owners, Engineers, and local residents.

An interesting item results in an examination of the age distribution of the more than 80,000 non-Federal dams that were in existence in 1985. It is noted that only about 15 percent of these dams are at least 50 years old; however, in just 15 years, by the year 2020, almost 85 percent will be 50 years old. These figures are pertinent since our experience gained during the Inspection Program underscores the fact that as dams age, they tend to outlive their intended usefulness, become susceptible to maintenance neglect, deteriorate structurally and become unsafe as design criteria or guidelines are improved.

The occasional failure of a dam illustrates that man-made structures do not last forever. Of the approximately 80,000 dams in the United States, some will fail regardless of whether inspections have been made, how well they have been built, and, yes, even if the dam's safety has recently been improved by rehabilitation. It is, of course, expected that the number of dam failures will be substantially reduced and I Believe we have experienced this development over the last few years.

Finally, to quote an old Chinese proverb that sums up dam safety issues and concerns best of all:

"When life knocks you down to your knees,
you are in a perfect position to pray"

Chapter 7

DAM SAFETY EVALUATION IN INDIA

Dr. Y.K. Murthy

Water Resources of India

Water Resources of India has been assessed as 1880 km3 on the basis of the total average annual run-off of the river systems of the country. It is interesting to note that the water Resources of India is almost equal to the estimated water resources of USA even though geographical area of India is about 40% of geographical area of USA.

The river systems of India has been classified into three groups based on the catchment areas of the basins, as follows:

Major Rivers:	River basins with Catchment Areas of 20,000 sq. km and above.
Medium Rivers:	River basins with Catchment Areas between 20,000 sq. km and 2,000 sq.km.
Minor Rivers:	River basins with Catchment Areas below 2000 sq. km.

The following are the percentages of total water resources contributed from the three groups of the river systems:

Major River Basins	84%
Medium River Basins	8%
Minor River Basins	8%

Thus India's 12 Major and 46 Medium Rivers contribute 92% of the Water Resources. Most of the river systems carry 80% of the flow during four to five monsoon months of the year. Hence the need for storage by construction of Dams.

Storages

Storages created by construction of Dams are as follows:

(a)	Up to 1947 (year of Independence)	13.65	km3
(b)	Up to 1985	143	km3
(c)	Under Construction	79	km3
(d)	Large and small tanks	30	km3
(e)	Underplanning	82	km3
	Total	334	km3

Irrigation and Hydro Power Development:

(a) **Irrigation potential:** Total estimated potential is 113 million ha of which 58 million ha is from Major and Medium Projects and 55 million ha is from Minor Schemes where a Major Irrigation Scheme has a culturable command area (CCA) of more than 10,000 ha; Medium Irrigation Scheme has a CCA between 2000 ha and 10,000 ha and Minor Irrigation scheme has a CCA less than 2000 ha. Present development of Irrigation in India is as follows:

Pre-independence (after partition):	22 million ha (10 million ha by canals + 12 M ha by Minor Schemes)
Developed Potential now:	32 M ha by Major Medium Schemes
	41 M ha by Minor Schemes
Total:	73 M ha
(b) **Hydro-power potential**	85.5 M kw at 60% L.F.
Pre-independence:	0.6 M kw (Installed capacity)
Developed Hydro-power now:	15.7 M kw (Installed capacity)

Thus, there is a continuous growth of Irrigation and Hydropower development after India became independent.

Dams

As of today, number of large dams in India as defined by ICOLD is 3,204 including most of the dams with heights between 10m to 15m, which satisfy at least one of the following criteria laid down by ICOLD:

1. the length of the crest not less than 500 m.

2. the capacity of the reservoir formed by the dam not less than 1 million cum.

3. the maximum flood discharge dealt with by the dam not less than 2,000 m 3/sec.

4. if the dam had especially difficult foundation problem, or

5. if the dam is of unusual design.

Safety aspects of World Bank funded Dams in India

I had the opportunity to collect the statistical data of all the World Bank funded dams and prepare on "Inventory of Dams Related to World Bank projects in India" in 1986 and later to assess the status of Dam Safety Assurance Programmes relating to World Bank Projects in India".

An analysis of the status of Dam Safety Assurance Programme for the 131 dams funded by the World Bank was carried out. The study showed the following results:

		Nos.	%
1.	Spillway capacity not satisfying Indian Standard Code (IS code)	20	15
2.	Freeboard not conforming to IS Code	25	20
3.	Seismic factor not taken into account in the design of the dam (mostly old dams)	15	11.5
4.	Non-study of Emergency Reservoir Operation Plans	90	70
5.	Distress Manifestations reported	36	27.5

The nature of distress manifestations were as follows:

1. Seepage and wetness at the toe/downstream slope of earth dam upon first filling and also during operation.

2. Seepage through the stone masonry dams during first filling and also during operation due to deterioration of cement mortar cracking in the old dams.

3. Scouring below the spillway bucket.

4. Alkali-Aggregate Reactivity was discovered in one of the concrete dams constructed in 1957.

5. Breaking of PVC/rubber seals at the block joints and leakage into the Drainage Gallery.

In addition, Dam safety Organization of central Water commission has recently collected data of similar distress on many more dams. In Maharashtra State alone, distress in 46 dams have been reported. It is seen that there are increasing number of dams showing some signs of distress, especially during first filling as well as in the older dams.

Dam Failures

ICOLD made a study of the failure and accidents to large dams. According to this study, up to 1965, out of about 9000 dams in the world at that time, 535 incidents to dams were recorded, of which 202 related to total

failure. In India, out of registered total 433 dams, at that time, 40
incidents have been reported including 13 total failures. Subsequently, there
have been some 7 failures like Machu Dam. The causes of such failures have
been investigated and suitable actions have been taken to restore these dams.
Machu Dam is being reconstructed and funded by the World Bank.

Failures of dams built to accepted state-of-art dam engineering have
also taken place in the world. It is interesting to note the observations of
Dr. Karl Terzaghi, professor at Harvard University, after the failure of
Malpasset Dam, quote: "In situation of this kind, it is at the outset,
impossible to divorce the technical aspects of the event from the human
tragedies involved. Yet every fair-minded engineer will remember that
failures of this kink are unfortunately, essential and inevitable links in the
chain of progress in the realm of engineering, because there are no other
means for detecting the limit to the validity of our concepts and procedures".

International Congress on Large Dams have been highlighting the safety
aspects of the dams since its inception and emphasizing proper design and also
quality control during construction besides proper maintenance during
operation after the completion of the construction.

Dam Safety Organization in Government of India

In India, under the constitutional dispensation of legislative powers,
the States have been assigned the primary role in the development of its water
resources. Hence all the dams are mostly owned by State Governments. A few
dams are also owned by State organizations like State Electricity Boards,
State Power Corporation, Municipal Corporations. Only about 6 dams are owned
by a private agency like Tata Electric Co. and those were constructed for
hydro-power generation in the 1920's.

Keeping in view the importance of dam safety in the country, a Dam
Safety Organization was established for the first time in May 1979, in the
Central Water Commission, Government of India to initiate and assist the State
Governments. This organization was given primarily an advisory role with the
responsibility for monitoring, taking remedial measures for mitigating the
distress and ensuring the safety of the dams resting with the State
Governments who own these structures.

The main functions of this organization were defined as follows:

1.   to document the salient design features of the dams in the country
      and compile the data on structural behavior of dams received from
      the States or other State agencies owning the dams;

2.   preparation of technical memoranda on dam safety issues and
      compilation of dam safety literature and articles;

3.   assist the State Governments to identify the causes of potential
      distress and redress them on specific requests form State
      Governments;

4.   collection and analysis of instrumentation data and preparation of structural behaviour reports for dams and hydraulic structures designed by CWC and also when received from other States.

In 1982, Government of India, constituted a Standing Committee to review the existing practices and to evolve unified procedure of dam safety for all dams in India, under the chairmanship of Chairman, Central Water Commission. The Standing Committee submitted the Report in 1986. The Report, not only gave the varying existing practices but also made specific recommendations for setting up Dam Safety Cells in the State Governments. Ministry of Water Resources of Government of India, accepted the recommendations of the Committee and requested the State Governments to enforce and implement the recommendations. A summary of the important recommendations are given below:

1.   All State Governments, who have significant number of Dams shall create a Dam Safety Cell/Organization, which shall be manned by competent technical staff and adequately funded.

2.   Dam Safety Organization in the State Government shall be created by the Irrigation Department, who incidentally own maximum number of dams. This organization shall also be responsible for the safety review of dams owned by other agencies in the States.

3.   It is necessary to prepare the completion report for each of the major dams with all the technical details and enumerate the problems encountered during construction and how they were resolved. This Report would serve as a guide for future dams.

4.   It is desirable that all the major and medium projects are guided by a panel of experts even from the stage of design, if not investigation, till the project is completed.

5.   Each dam should have an operation and maintenance manual, which would contain detailed instructions, procedures and rules of operation and maintenance of the dam.

6.   Adequate flood-forecasting system with wireless communications need to be established for reservoir operation, where such system is likely to help in giving advance information about incoming floods.

7.   As it may not be possible to prepare emergency preparedness plans for all the dams immediately, they shall be prepared for all the dams after fixing priorities, depending on the hazard potential.

8.   All State Governments shall arrange safety reviews of dams which are more than 15 meters in height or which store more than 50,000 acre feet of water by an independent panel of experts once in 10 years.

9.   Legislation, at present, may be confined to a broad framework and guidelines covering salient recommendations of the Standing Committee discussed above.

In 1987, Government of India reconstituted the Standing Committee as the "National Committee on Dam Safety". The terms of reference to the Committee are to:

1. monitor the follow-up action on the Report of Dam Safety Procedures prepared by the Standing Committee both at the Center and at the State Government levels.

2. oversee dam safety activities in various States and suggest improvements to bring dam safety practices in line with the latest state-of-art consistent with Indian conditions.

3. act as a forum for exchange of views on techniques adopted for remedial measures to relieve distress, developed in the dams.

All the State Governments having significant number of dams are represented in the National Committee on Dam Safety. The National Committee will meet twice a year and the Dam Safety Organization of Central Water Commission would act as the Secretariat to the National Committee and coordinate the activities of the State Governments regarding dam safety aspects.

Dam Safety Cells in State Governments

With the initiative of the Dam Safety Organization of Central Water Commission, Dam Safety Cells have been established in 12 States where there are considerable number of dams under construction and operation. The total number of dams located in these 12 States is about 3100. A cursory look a these organization show that except for two or three States, the organization is neither satisfactory nor has been staffed with qualified and experienced personnel, though sincere efforts are being made by the Dam Safety Organization of Central Water Commission. In other States, wherever, it is not possible to set up such cells, especially where the number of dams in the State does not justify, Dam Safety Organization of Central Water Commission, would be strengthened for assisting those States Governments.

Dam Safety Publications and Seminars

1. Dam Safety Organization of Central Water Commission has prepared the following documents and circulated to the State Governments.

2. Guidelines for Safety Inspection of Dams

3. Guide lines for Emergency Action Planning

4. Data Book Format

5. Proforma for Periodical Inspection of Dams

6. Suggested Check List for Inspection of Dams

7. Modes and Causes of Dam Failure

8. Potential distress problems affecting the safety of dams -- Causes and Remedial Measures

9. Format for collection of Data on Dams from State Governments

Seminars and Workshops

Dam Safety Organization of Central Water Commission had also organized Seminars and Workshops, sometimes at the Dam Sites, to discuss various aspects of Dam Safety, when all the concerned engineers were invited to contribute papers and participate in the discussions. From October 1983 up to now, eight Workshops had been arranged. Two more symposiums are programmed for the current year.

As can be seen from the information given above, activities of the Dam Safety Organizations in the Central Government as well as State Governments have been primarily focussed on the existing dams, though one of the functions listed includes Review of quality control data for new projects. However, National Committee has recommended desirability of constituting a Panel of experts for all the major and medium projects even from the stage of design, if not investigation, till the project is completed.

UNDP Participation in Dam Safety Activities

A UNDP Project titled "Updating the Design, Instrumentation and Surveillance Technology for Dams and other Hydraulic Structures" has started its operation from March 1987 for the Designs and Research Wing of Central Water Commission. Under this Program twenty engineers had been deputed under fellowship training for acquiring new technologies in various fields below:

1. **Instrumentation Technology**

2. **Dam Safety Surveillance**

3. **Finite Element Analysis**

4. **High Earth And Rockfill Dams, and**

5. **Design of High Head Gates**

UN assistance is also obtained for procurement of equipment like Photogrammatic Camera, Underwater TV Camera, Remote Operated Underwater Vehicle for scanning the submerged face of the dam, Scuba Diving sets, and Software Programmes for design and instrument data processing.

Under this UNDP Project, experts from other countries in the field of Dam Surveillance, Instrumentation Technology are being deputed for short-term to assist the Dam Safety and Design Organization of CWC. So far, two experts in the field of High Earth and Rockfill dams had visited the Organization.

Role of World Bank in Dam Safety Assurance Programmes in India

As indicated earlier, the World Bank has funded about 130 dams either directly or funding the downstream irrigated command areas coming under existing dams. As such, the World Bank has been keenly interested in the dam safety aspects of not only the existing dams but also the new dams proposed for construction under the World Bank projects.

Existing Dams

In the existing dams, there are two categories, namely the newly constructed dams ready for first filling of the reservoir and the old dams which have been in existence and serving eh needs of irrigation, hydro-power etc. Both these categories of dams require close surveillance.

In the first filling of a large reservoirs behind a newly constructed dam require vigilance. Distress manifested should be closely watched by a team of experienced dam engineers and engineering geologists. Seepage at the toe of the earth dam, leakage through the body of stone masonry dam, disturbance on the stone pitching of the upstream slope of the earth dam, leakage through the abutment hill etc are some of common distress manifestation. Remedial measures are carried out and such distress manifestations invariably cease after a lapse of time. Thus these distress manifestations during first filling of the reservoir are taken care of.

In the case of old dams, thought the causes of distress have been identified, by dam review panels, remedial measures have not been taken up. World Bank Supervision missions have been drawing the attention of the respective State Government Organization, repeatedly. Thought the Governments have recognized the need to carry out remedial measures, because of paucity of funds, the implementation has been delayed. Some of these old dams are really in critical condition, which need immediate attention.

Providing additional spillway capacity on the basis of the revised hydrological studies to conform to the Indian Standard Code has been one of the major items, which is not implemented for many of the existing dams, apart from other remedial measures.

New Dams

With the initiative and persuasion of the World Bank, State Governments have agreed to constitute Dam Review Panels (DRP) for the Dams included in the New projects funded by the World Bank. DRP's are in operation in almost all the new projects. The World Bank has identified some deficiencies in the effective working of these panels and has recently prepared a revised set of terms of reference in consultation with Government of India. Some of the deficiencies that were found are as below:

1.   Number of members in the panel of the required specialist disciplines to fully reflect the needs of the Project.

2.   Selection procedures to ensure highly experienced, professionally active specialists.

3.   Venue and frequency of the panel meetings.

4.   Degree of independence on operation and specialist support to the Review panels. In case the latest techniques are not available, provision for including foreign experts either as permanent or occasional members of the panelist, is not easily accepted in spite of DRP's specific recommendations. There seems to be some improvement in this aspect in the recent days.

5.   Effective interaction with the respective design consultants and construction organization.

6.   More effective mechanism to monitor the implementation and compliance of DRP recommendations.

In order to bring out the experiences, and highlight the deficiencies of Dam Safety Assurance Programmes currently being followed, a joint dam safety workshop was conducted between the World Bank and Central Water Commission at New Delhi on March 14-15, 1988. As a part of conclusion after the deliberations of this workshop, recognizing the fact that CWC has the responsibility only for establishing standards and guidelines and has no authority to ensure enforcement, a review of the procedures in vogue at the State level would be an essential step towards understanding the present situation. Accordingly a joint review by CWC, the World Bank and State representatives was undertaken in Maharashtra State where more than half of India's large dams are located, in December 1988.

Summary of broad conclusions and recommendations for modifications and improvements in the functions of DSO of Maharashtra Government are given on next page:

1.   The efforts made by Government of Maharashtra for establishing a Dam Safety Organization to carry out efficiently the periodical inspections of existing dams have been satisfactory, though it was found that the organization is understaffed and needs to be strengthened adequately to enable more intensive involvement in the implementation of dam safety procedures.

2.   There is need to make the framework of dam safety assurance, as separate organization, which should have a strong degree of independence from design, construction and operation organizations and also authority to take positive actions in the implementation of remedial measures required to ensure the safety of the dam.

3. Associating the DSO, even from the initial stages of planning and construction of new dams, to enable establishing a strong and independent in house Dam Safety Organization for taking care of dam safety aspects at all stages and gradually eliminate the need for independent dam safety review panels in the long run. Though this proposal was accepted in principle, in view of the large number of existing dams requiring special attention, State Government agreed that this would be considered of implementation in a phased manner after completing the work on the existing dams.

4. Dam Review Panels have only been established for projects financed by the World Bank. There are no such panels for other projects. Independent Review Panels are essential for other projects. In addition, it was observed that the staff engineers of the Design organization of the State Government visit the construction sites normally once a year. Regular site visits should be at much shorter intervals. Dam Review Panels should cover all the major and medium projects and should be established as a temporary solution until their functions can be undertaken by the strengthened Dam Safety Organization, which is expected to be done in the third phase of development plan of DSO.

5. With some modifications and improvements, the team found that the framework of Dam Safety Organization in Maharashtra can be developed as a model for implementation in other states. It was estimated that a total period of about eight years is deemed necessary to develop the existing DSO into a strong and independent unit to handle all dam safety aspects from planning to final operation stage.

Conclusions

India has made rapid progress in establishing a Dam Safety Organization at the Central Water Commission, Government of India, which has been striving hard to impress upon the State Governments, who are responsible for planning, construction and operation of dams, to establish a strong dam safety organization in each state. It has succeeded in its efforts to set up Dam Safety cells in 12 States, where there are large number of dams. But these organizations are far from satisfactory except a few States like Maharashtra, Gujarat. These organizations should be strengthened and staffed with adequately trained personnel to be able to evaluate the dam safety aspects from planning to final operation stage. More importantly, there should be a more effective mechanism to monitor the implementation and compliance of the findings of such organization and adequate funds should be provided.

It is heartening to note that the Government of India has constituted a National Committee on Dam Safety in October 1987, with the primary functions of overseeing the dam safety activities in various States and also to advise ont he improvements to bring the dams safety practices in line with the latest state-of-art consistent with Indian conditions.

Furthermore, this National Committee has created an intense awareness of Dam Safety. It is evident from the fact that the representatives of State Governments brought several case histories of dam distress in the initial meetings of this Committee.

Dam Review Panels are expected to be established for all the major and medium projects even from the stage of design, if not investigation, till the project is completed in accordance with the recommendation of the National Committee on Dam Safety.

With continued discussions and interaction with The World Bank, Government of India hopes to develop the Dam Safety evaluation programmes to cover all the major and medium dams in the country covering not only the existing dams but also the new dams, which would be undertaken for construction in the future.

Chapter 8

DESIGNING SAFETY INTO DAMS

P.A.A. Back

My subject is designing safety into dams and to put this into the context of some of the problems we face I would like to refer to the case history of the Manchu Dam - II in India. The initial design flood capacity of the spillway was 200,000 cusecs based on an estimated PMF of 191,000 cusecs. However in August 1979 the dam was overtopped by a flood of 460,000 cusecs. As a result of this a complete review of the hydrology lead to a revised PMF estimated at around 739,000 cusecs, which is nearly four times the original design flood. Even as this revised design was about to be built a still greater flood occurred which required the PMF to be raised still further to around 933,000 cusecs. This is five fold increase on the original design flood in just 20 years.

If nothing else this illustrates the levels of uncertainty with which we have to deal in the design of dams. It is therefore ludicrous to put precise figures to some of our assumptions or to believe it possible to define limiting conditions precisely.

Foundation conditions are the other major area of imprecision. Even the most exhaustive geological exploration of the site may still fail to yield all that is below the surface which is relevant to the design of the dam. Thus it is essential that margins for contingency are substantial and our approach broad and bracketing - not fine and finicky.

It is in this context that I want to examine a fairly new phenomenon that is becoming fashionable in some circles on the matter of dam safety - and indeed not only dam safety but engineering structures generally.

This approach is known as Economic risk analysis, based on probability theory. The idea is that it is possible by rational means to define the level of investment which it is appropriate to apply to, say the design of a spillway, so that in the event of failure, the replacement cost plus the cost of all damages is a minimum.

Personally, I find it very hard to take such approaches seriously.

First it requires the definition of the probability function of the particular risk, something which is itself fraught with much uncertainty. It requires the fixing of a discount rate which must apply over the life of the dam. It requires an estimate of the cost of damage from a devastating flood wave released downstream. Finally it requires that a value be placed on the human lives which would be lost in such an event.

There is no way that the public would find it acceptable for such a clinical and detached approach to the safety of dams to be taken. Those who urge such approaches do nothing but bring the whole subject into disrepute.

Number crunching and statistics do not, and cannot, solve all the problems we have to address - though, of course, we do depend upon them to guide us in many areas of our work.

No - dam safety must be seen to be the dominant consideration of what we design taking into account all factors that can reasonably be identified - and don't let us try and define what is "reasonable". The safety of dams will depend on 3 main factors, whatever else we may choose to consider in such fields as risk analysis, statistics, or other esoteric disciplines.

First, it will depend on the design. - How well has the designer understood the very many different facets involved; - The foundation geology, the floods to be routed, the materials to be used in construction, the durability of those materials in the given environment.

Secondly, safety will depend on the quality of construction:

Does the quality of construction match the assumptions of the designer - or has sloppy workmanship invalidated the design assumptions. Have poor materials been used in place of those specified.

Thirdly, safety will depend on subsequent operation and maintenance of the dam. There is no such thing as a "walk away" dam - one that can be left to its own devices. It must always remain under supervision - it must always continue to be monitored.

Each of the above factors - design, quality of construction, operation and maintenance, are essential elements which go to determine whether or not a dam is safe. Remove any one of them and you have a potential catastrophe. Today I want to look in particular at the first of these factors - <u>Design</u>.

How Do We Design Safety Into Dams?

I suggest there are certain very broad, but relevant approaches.

Recognize the Limitations of the Design Assumptions and the Analysis

It is astonishing what authority seems vested in the printed word - especially if it is in computer print out. We all say to one another "Don't believe it just because it has come out of a computer" - then promptly do just that.

It is also astonishing that designers will sometimes go to great lengths to respect certain theoretical criteria such as maximum permissible stresses - yet the assumptions which have been used in the analysis may be breathtaking in their crudeness. For this reason it seems to me far better to adopt what I would call a bracketing approach. Accept the fact that many of our assumptions, such as foundation modulus may be inaccurate or downright wrong. The analysis should look at a spectrum of moduli, not just one. The designer must <u>frame</u> the problem before he can truly solve it.

Wherever possible have at least two lines of defence. Design redundancy into the structure. That is not a waste of money - it can often be achieved at very little extra cost. Take a concrete gravity dam. If it is my responsibility to design such a dam I will always look for a way to build it on a gentle curve - never on a straight axis. A gentle curve greatly increases sliding resistance of any one dam block at almost no extra cost to the structure.

Yet most existing gravity dams, or under construction, have a straight axis. I would also introduce a gentle curve for rollcrete construction - yet here again one can note that a straight axis is very often adopted (Willow Creek, Upper Stilwater).

Redundancy in spillways is also very desirable so that if for any reason in spillway is out of commission at least there is an alternative, even though it would not on its own be able to handle the design flood.

Be very careful with the concept of a factor of safety. Take an arch dam for example. We used to depend heavily on model tests to prove the adequacy of such dams. Indeed clients frequently used to insist that models to a certain scale were constructed. These model tests would be used to carefully measure the compressive and tensile stresses induced in the arch.

Then finally we would test to destruction and often find that loads up to 5 or 6 times the normal maximum could be sustained by the arch before failure. And some would argue that means a factor of safety of 5 or 6. In fact it means nothing of the kind. In the first place the type of load above normal maximum cannot occur in nature - except in a case such as at Vaijant when a 300 ft. wave passed over the top of the dam so the loading assumptions in the overload tests are fanciful in the extreme and the appropriate pore water pressures within the foundations are ignored completely.

Indeed it is almost certain that if failure occurred it would be in the foundations--the one area we cannot model exactly--the one area where our knowledge is least. Our laboratory model test is at best half a test - and the most important half is missing.

Be very careful to differentiate between stabilizing conditions which act in parallel or in series. To take a simple example, consider a dam block on its foundations with its downstream end face abutting against the excavated rock. Sliding failure of the block is resisted by two mechanisms. Shear resistance of the concrete/rock interface and the passive resistance of the downstream mass of rock against a downstream movement of the dam block.

The sheer resistance on the foundation plus the passive resistance of downstream rock mass may together greatly exceed the destabilizing forces acting in the dam block. Yet the block could still fail - because in fact the resistance mechanisms don't act together - in parallel. The passive resistance is only mobilized once the first shear resistance has failed. So the defenses are overcome one by one. This is what I call "tearing along the dotted line"; and it is surprising how often the incautious designer can be caught out this way.

Let me illustrate with another example

Assume we have a dam built on a foundation consisting of hard rock interspersed with zones of softer rock. As the load is applied the softer zones yield without helping to carry any of the load which is therefore all transferred to the hard rock. As the process continues the load concentrates onto the hard rock which may become overloaded. We then have the paradoxical result that failure occurs at the strongest point.

It was exactly this consideration that we had to deal with on the Kariba dam. There we had good solid gneiss over most of the foundation except in the top half of the right abutment. Because this zone could not absorb the load from the arch, but would merely compress under load, we worried that the lower good rock which would then have to take the load, might become overstressed and fail.

Faced with this problem we decided to build 4 massive buttresses, which would carry the arch load through the compressible rock and onto sound solid rock at a lower level. In doing this, we then, of course, had to ignore any contribution to safety from the softer rock. It could not make any contribution with the more rigid buttresses in place.

Be careful of mechanisms which bypass the defenses. Sometimes it is a small matter of detail yet it can be very important to the success of a design.

Don't depend on high tech, sophisticated devices or computerized electronic equipment for the safety of a dam. This comment is particularly relevant in a Third World context.

For example, the use of computerized flood routing has in one case I know of led to tragedy. In this case malfunctioning of the equipment led to the spillway gates opening suddenly and drowning a whole family. It could equally well happen that when the gates were required to open nothing would happen and the dam be overtopped. In tropical conditions the deterioration of electronic equipment can be rapid and realization of its disrepair may come too late.

If we look at the history of dam failures there is a considerable preponderance of failures due to overtopping, especially of embankment dams. Overtopping has generally occurred because the gate operator failed to react in time to the arrival of a flood, or power was lost at the critical moment and the gates couldn't be opened.

So why don't we build dams with a free ungated overspill with no gates. Certainly that is one option we should always consider - and that in fact is what we are planning for the Katse high dams in Lesotho. But sometimes that is a very expensive option in terms of loss of available storage.

For example, the Victoria dam in Sri Lanka--here the economic evaluation of the scheme demonstrated very clearly that to maximize benefits we had to construct the dam to impound water to the highest possible level--constrained only by not flooding the town of Kandy some 16 km upstream. If we adopted a free overspill the normal top retention level would have had to be some 8 m

lower than with a gated overspill to allow for the flood surcharge. So we
really had to have gates--but we were most concerned that because with gates
the flood surcharge would be very small, the gates had to function correctly-
-they had to open in time without fail.

Indeed, as I was considering this problem in my hotel room in Kandy,
there were no less than 3 power cuts as a storm raged outside. It is in just
such circumstances that power failure to the gate operating motors is most
likely to happen--and when it is most important that the gates can open.
So it was that I felt we had to come up with an answer--a gate which would
open without fail at he right time and by the right amount and totally without
dependence on power or a gate operator. And that indeed is what we now have
on the Victoria dam and what we are also installing on the Samanalawewa dam in
Sri Lanka. So much for some of the design considerations but as I indicated
earlier there are also construction and maintenance angles which also impact
greatly on dam safety.

At the end of the day, whether a dam stands or falls, I suggest is not
likely to depend on the sophistication of the risk analysis. But it most
certainly can depend on what happened at site, late on a Friday evening in the
pouring rain when the work force were about to stop for a long. weekend.

So what can we do to ensure that dam safety is truly realized?
I would suggest, that just as there are the three facets--design,
construction, maintenance--so are there three areas of activity which dam
safety must address--and perhaps therefore three distinct but overlapping
inputs required.

And if I may say so, I believe that you as an interested party in the
financing of the construction of a dam should have well defined requirements
in each of the three areas.

Design

Here it is already the practice to have an independent Panel to review
the design of the project consultants: I believe this is good - but its value
is very dependent upon the quality of the Panel and how much time they can
devote to looking, not only at the broad concepts of the design - but at the
details as well. Sometimes it is in the detail that designers have gone badly
astray. So I believe it is important not to skimp on the Panel's input. And
the input should be as early as possible to avoid major problems if the Panel
wants things changed.

This problem arose on the Magat dam in the Philippines. I was one of a
Panel which was formed only after the project consultants had already gone a
long way to drawing up the Tender documents. We disagreed fundamentally with
what was proposed - and as you can imagine it led to considerable resistance
to making the changes we wanted. In the end they were made, but much of the
trouble would have been avoided if the Panel had been involved from the start.

Construction

In many ways this is the most difficult aspect of all. In my experience
the level and extent of site supervision varies widely from country to country

and site to site. Many clients in the Third world resent having to pay for a large supervising team and believe them to be superfluous.
They also argue that especially at the level of inspector where formal qualifications are rare - the work could equally well be performed by their own people.

These arguments must I believe be resisted. And I believe it is right that the Bank should insist on appropriate levels of supervision of construction of a project if they are involved in its funding. What the level of supervision of construction should be is something the Review Panel should also advise on.

Maintenance and Operation

Here too, I believe the Bank has an important role t play - to ensure that the owners set up an appropriate organization to undertake routine inspection and maintenance and for the Project Consultants to continue to have an ongoing commitment to inspect and advise on the safety of the project. It is very important, that those who have the most intimate knowledge of a project--those involved in its design and have supervised its construction, should continue to monitor and comment on its subsequent behavior. It should be foreseen and budgeted for from the very outset that such would be the project consultants duties and obligations. Unfortunately in many cases the matter is not considered at all until the project has been built and then either nothing is done--or at best ad hoc arrangements are made--sometimes only because a crisis has arisen.

Chapter 9

REASSESSMENT OF SAFETY OF EXISTING DAMS

D. Bonazzi

The world has few examples to show of reservoirs which have been emptied as to return the land to its initial fallow, pastoral, or cultivated state. Once in place, and irrespective of its purpose, an artificial reservoir generally becomes a practically permanent part of the landscape and environment which it has usually profoundly modified.

From time to time, and in special cases, it is thought of eliminating a reservoir, or just draining it temporarily, but at the same time there is opposition to this movement, and real difficulties of practical impossibility can bar the way. Examples of this are those reservoir which are so gigantic that no acceptable dewatering works could be included in the projects for towns which have grown too quickly.

Because of this debate, specialists of all sorts in matters concerning dams have an ever increasing role in the monitoring, maintenance, and repair of more or less old dams, as well as their involvement in the design of new dams. In fact, existing dams now account for about 10% of the dam-related work of Coyne et Bellier.

The examination of existing dams should by no means be taken to be a minor activity in relation to the design of new structures, for the dame high degree of responsibility is involved. The problems faced are often similar to those encountered on undeveloped sites, but in addition there are those resulting from time effects. And time affects both materials and human memory.

No illness can be treated before it is identified. This also applies to existing dams, which are complex structures designed, built, and operated by distant or disappeared generations who have left their mark and passed on a more or less mysterious heritage. A solid historical knowledge of construction techniques is then most valuable, for, as an example, although practically no masonry dams have been built for very many years (at least not in developed countries), the specialist may be required to know what to expect of such a dam when its performance is assessed. Even so, a simple survey of the structure can be difficult; and it can be alarming.

Investigation of the Dam and its Foundations

Records

The best possible case is when monitoring results are available. Most frequently such records are of topographical monitoring, and may or not be reliable as a result of movement of the survey monuments and unresolved contradictions.

Even the design drawings are to be regarded with caution. Engineers are often poor archivists, and the result can be an unorganized collection of drawings of design alternatives which were not actually built, final design drawings (whose representation of the actual structure "as-built" is in any case often to be mistrusted), and any number of other suspect documents. This has even been the case for dams less than 20 years old. The engineer need therefore be patient and circumspect in examining the available information. He can then look directly at the dam itself.

Visual Inspection

Visual inspection is important, for the engineer can draw up a list of his own observations such as cracks, leakage, mortar deterioration, deformation, scour, vegetation (sign of seepage), etc., and traces of past rehabilitation and repair can sometimes be seen as well. With dam records, and sometimes with the memories of past operators, the history of the dam can be reconstituted. This covers such events as the heaviest floods experienced, the results of nay drawdowns, known sedimentation, etc.

It is becoming more and more frequent for the submerged parts of dams to be inspected by divers or by remote-control video cameras. But there still remain a great many difficulties, such as for example the pin-pointing of the zones examined, the scale of images received and associated lighting problems (at the distance required for a view of a large portion of the dam wall there is insufficient visibility).

Investigations

Investigations can be carried out directly on the materials of the dam and its foundations. Sometimes the foundation drawings are not reliable and boreholes have to locate the dam/foundation interface.

Dam Materials

The materials of which the dam is built can be analyzed in a laboratory using samples of concrete, stone, mortar, or fill. Geophysical tests can also be carried out on the site. These can include permeability tests (especially in boreholes) and piezometric observations. Cracked concrete is usually given very special attention in the form of boreholes with borescope inspection and dye tests, etc.

Foundation

Frequently little or no serious geological investigation has been performed on the foundations of the older dams. In such cases there is a minimum amount of investigation which should be done, for which the assistance of a geologist is as indispensable as for a new dam. The geologist's work is not easy however, for there are no longer any visible excavations, surface rock is weathered, and the reservoir can be full. But good geological data of a general nature may have become available since construction; in any case, investigation holes must be drilled if necessary. Geophysical prospecting techniques can also be a good means of investigation.

Hydrology

Hydrological data must be checked in all cases. It often leads to the conclusion that the flood risk was underestimated, which means that the dam owes its continued existence to the fact that the heaviest floods it has actually encountered are not really the heaviest possible. The rehabilitation of the dam then calls for the spillways to be brought up to modern standards.

Stability Analysis

When it is decided that works must be done on a dam, it is generally because certain obvious weaknesses have been revealed. Such disorders can include visible deterioration such as cracks or settlement, or even certain design errors which were unfortunately frequent at one time, including gravity dams whose areal cross-section is insufficient according to modern criteria, or insufficient spillway discharge, etc.

The engineer will obviously try to apply modern safety criteria. In his calculations he will use the materials properties determined by the investigations mentioned above. And of course the means of calculation available to him are far more powerful and far-reaching than what the designers used: sophisticated modern numerical models can even analyze the strength of cracked concrete structures.

There are still major difficulties though:

1. assessing the mechanical strength of disused materials such as masonry; and the problem is aggravated by the fact that the materials in place may have aged;

2. assessing the behavior of fill which may well not be particularly homogeneous;

3. determining the coefficients of friction and cohesion for cracks;

4. determining maximum pore pressure in fill;

5. extrapolating data measured on the dam now to determine the figures which could be obtained under extreme circumstances (floods for example).

One should not underestimate the importance the assessment margin has on parameters of this type. In fact, the margins are sometimes such that using the minimum values (conservative) does not explain how the works continue to stand, and the engineer will avoid making a fool of himself by showing that, theoretically, the structure cannot withstand its stresses when in fact it does, and may in some cases have been doing so for a century. But it is certainly true that some masonry gravity dams built around the turn of the century have a cross-section which is today judged insufficient, a radius of curvature os slight that it does not appreciably add to the safety of the structure, and no drainage system. All that can be said in these cases is that the dams, in their actual condition, do indeed have a certain safety factor which is demonstrated by their very existence, but that it is slight and cannot be precisely evaluated.

The same sort of thing applies to other types of dam as well.

And for all types, stability under earthquake loading may also have to be dealt with. As for floods, existing dams may well have experienced no earthquakes, but that does not mean that the earthquake risk is nil. A special assessment is therefore necessary and sometimes calls for additional reinforcement. There has obviously been a great deal of progress in the identification of seismic zones in recent years. It sometimes happens that the conclusions of stability analysis are so disastrous and the only possible means of improvement so uncertain that the dam is declared "unfit for continued operation: to be replaced".

Typical Disorders

Concrete and Masonry Dams

Practically all disorders in concrete dams are due to concrete volume change. Concrete is a fragile material, and generalized or localized volume change sets up tensile stresses which cause cracking. The end result is leakage and associated complications. Some concrete dams are damaged by deformation of their abutments (Zeuzier, Switzerland), but this is a far less frequent occurrence than concrete swelling or shrinkage.

There are multiple causes of volume change in concrete. Firstly there is temperature, which varies to different degrees depending on the climate, and which will have a greater effect on lighter structures such as arch and buttress dams than on heavier gravity dams. Evolution within the concrete also causes swelling and shrinkage. Swelling of the concrete itself is a physico-chemical phenomenon involving the composition of the cement and the type of aggregate (alkali reaction). Shrinkage can be due to creep under loading or shrinkage prolonged beyond what is normal for the curing time. The table overleaf gives some idea of the scale of volume change in some large dams.

Concrete swelling produces the cracking along the toe of arch dams which is the direct result of upstream tilting. Shrinkage makes them lean downstream (the dam blocks rotate about their foundation line), and if the grout curtain beneath the dam is placed too far upstream, dam and curtain lose contact, thus allowing reservoir water and pressure to move further beneath the dam.

Every effort must also be made to prevent reservoir water entering any cracks in the dam itself. Unfortunately the perfect membrane which can be glued to the upstream face of a dam to bridge cracks of varying (and variable) widths (up to 2 or 3 mm) has yet to be invented. For gravity dams, the facing invented by Maurice Levy (small arches isolating the upstream face from the dam) is an excellent but costly solution.

Modern techniques using epoxy resins can return a cracked concrete dam to its monolithic state (Zeuzier), but it must be considered that some cracks are in fact hinges which must be free to open and close: they must not be grouted up. Mechanical reinforcement of thin arch dams is difficult, for it is not easy to combine them with other structures which could share their loading. At Tolla, the original thin arch now retains only its role of watertight barrier, and the structure added on the downstream side is capable of taking the full thrust of the reservoir.

Concrete or masonry gravity dams can be reinforced by throwing an embankment against the downstream side, and this is a solution which is rapidly gaining favor. It has been used, for example, at Jorf Torba in Algeria and Joux in France. Since the lime mortars used in old masonry dams will not stand up to high hydraulic gradients for extended periods, they cannot be drained exhaustively, and it is the downstream embankment which must act as drain. Rockfill is preferred for this purpose.

In 1976, a 62.5 m high Indian masonry dam, Talakalele, was reinforced with an earth embankment on the downstream side after several attempts at grouting the masonry. Since the embankment is now a bed of high pore pressures and the leakage rate is still increasing (170 1/s at the moment), it was decided to engage further studies and rehabilitation works. Coyne et Bellier has recently signed a contract for this delicate rehabilitation which must necessarily be done without draining the reservoir. A great deal of work by divers is anticipated. Vertical pre-stressing of gravity dams was used on several occasions by Andre' Coyne in dam-raising projects, but this technique has lost favor because of the uncertain lifetime of the tendons (rust). It is sometimes used as a last resort.

Embankment Dams

Piping. Apart from the risk of scour of the embankment by overspill if spillway capacity is not sufficient, the main risk facing embankment dams is that of retrogressive erosion which can lead to piping. To know whether or not an old embankment dam is susceptible to this hazard requires i) knowing how erodible the fill material is (granular analysis, pinhole tests on clays), and ii) checking water levels and leakage (location, discharge, entrained solids) to see if internal percolation is stable and suitably located. Special attention should be given to the interfaces between fill and rigid members like culverts, for these interfaces are often the starting point of piping. A very old dam in Ceylon failed in this way a few years ago. Another case of collapse due to culverts is Wadi Qatarrah in Libya. Dam failure due to piping is reputed to be fast, and has indeed been so in many cases, but there are also cases where visible sand deposits at the toe of an embankment dam have shown in good time that piping had begun, even if only during high-water periods. Preventive measures could then be taken (Hijuro dam in Argentina for example).

Slides and Other Damage Resulting from Excessive Leakage. It is always possible to reinforce embankments by placing draining embankments ont he downstream side and, though more rarely, on the upstream side (as at Merdja dam in Algeria). When the problem is one of lost internal impermeability (cracking, internal erosion), any difficulties arise chiefly because of the ancillary works (Jatiluhur), for the diaphragm wall technique is now mastered (Balder Head in Great Britain, Fontenelle in the USA). It is expensive however. The reduced watertightness of Grobois II dam in France was improve by placing clay on the upstream slope.

Rockfill Dams with Impermeable Upstream Facings. Rockfill dams with impermeable upstream facings can withstand high rates of leakage without compromising safety. Even if it is a digression from the matter in hand, it is interesting to look at some spectacular examples of repair of such dams: Paradela dam (Portugal) and Rouchain dam (France), both rockfill dams with reinforced-concrete upstream facings, were successfully made watertight (after draining the reservoir) with a geotextile impregnated with an emulsified asphalt and a polymer. But there have also been cases of failure with sealing membranes on old dams.

Foundation

Foundations must be examined from a variety of viewpoints.

Watertightness. Special sealing works justified by the value of the water alone might have to be undertaken if leakage is high. This is not our concern here though if it does not compromise dam safety.

Stability. Foundation instability manifests itself in the form of slides or as piping. The solution is practically always to improve percolation flow, and generally involves one or both of the following:

1. reinforce or even rebuild the grout curtain,

2. clear the existing drains or drill new ones fitted with graded filter material if necessary.

Arch dam foundations can give rise to stability problems when the concrete has continued to shrink long after grouting up of the blocks. Consolidation grouting and even underpinning are also possible (Bouzey after its first failure).

Slides into the Reservoir

There are complex cases where slides into the reservoir start near the dam, threatening works such as tunnel portals etc. An example is Bou Hanifia in Algeria, where the problem is currently being examined, and Derbendi Khan in Iraq, on which Coyne et Bellier worked between 1976 and 1985.

Trends

Very good piezometric conditions have been found in some old dams when the original arrangements have been fortified by a watertight apron of fine sediment (alluvium) on the upstream side of the dam (Mohaned V in Morocco). This phenomenon is rare however. The general trend is for the effect of reservoir pressures to gradually move downstream through the dam and foundation as the years go by. This then calls for a counterattack to push the pressures back upstream, in so far as it is possible. The two conventional means for doing this are:

1. grouting upstream

2. drainage downstream.

The classic difficulty is that these works must sometimes be carried out under full reservoir pressure, simply because it is impossible to empty it. This can mean that the grout does not set in the place where it is injected and is carried further downstream and into the drains, thus blocking them further. Special steps must be taken to prevent this.

Lack of Monitoring Equipment

When dealing with existing dams, one is often confronted with a total or partial absence of instrumentation. Suitable equipment must therefore be installed before a reliable diagnosis can be given. The most important equipment includes:

1. flow meters

2. piezometers in the dam and its foundation

3. settlement gauges, and pendulums if possible.

The same instruments will be used for monitoring after the consolidation of the dam.

Consultant's Responsibility

A consultant who undertakes to reassess the safety of an existing dam cannot be held responsible for the safety of the structure the very day he signs the contract. Nor can he sign a contract without having previously carefully examined the problems at hand. However, this examination can but slightly reduce the time between contract signature and the moment when full responsibility for the works becomes his. In short, the transfer of responsibility takes time.

Before calling on a consultant, it is the project owner who is primarily responsible for the dam. Once a consultant is involved in reassessing the safety of the works, responsibility is gradually moved onto his shoulders, irrespective of the weakness and shortcomings which might come to light in the design, maintenance, documentation and physical state of the works.

Of course, the terms and conditions of the contract are important, but the sharing and handing-over of responsibility are difficult to define. It is also difficult to determine just how long a period is required for the transfer of responsibility to take place. And in the event of problems, only experienced professionals can judge whether the right procedures were engaged with all necessary haste and discernment. It goes without saying that the consultant can only be held responsible if the Owner has given him ample freedom to undertake all necessary investigations and, once informed of the consultant's views, has acted on them.

The fastest and least expensive manner in which to improve or re-establish an adequate degree of dam safety is to lower the reservoir level. This is sometimes a permanent solution when the purpose and arrangement of the dam are suitable. But is most often a good interim solution which gives all those involved to do their job calmly and efficiently.

Chapter 10

EXPERIENCE OF THE WORLD BANK IN IMPLEMENTATION OF DAM SAFETY

P.N. Gupta

The Bank has always been concerned with and given top priority to safety of dams on the Bank-financed dam projects. The Bank normally requires that experienced and competent engineers, acceptable to it, be responsible for the design of the dam and the supervision of its construction. The Bank seeks to promote all possible measures and coordinated efforts for:

1. preparation of technically sound and economically viable dam projects;

2. safe and successful implementation of Bank project;

3. safe monitoring of completed dam projects.

Bank Guidelines for Safety of Dams

Being very much concerned for the safety of Bank-financed dam projects, and to achieve the objective of safe and successful implementation of Bank provided specific Guidelines for Safety of Dams (#3.80, 1977) and insists on the following specific requirements:

1. experienced/competent engineers for design and construction supervision of dam;

2. independent Panel of Experts (POE) for review of the concept and designs/construction;

3. periodic inspections of dam after construction by qualified independent experts.

Under item 2 (1) above, "Competent Engineers," it is essential that the Project Consultants are selected in accordance with Bank Guidelines for the use of Consultants by World Bank Borrowers and by the World Bank as Executing Agency. It is also important that the Engineering Study of a dam project should not be parcelled or divided into pieces to suit the Bilateral co-financing for the project study, and should be completed as an overall single study so as to necessarily maintain the responsibility, homogeneity, reliability and technical direction by one overall engineering consultant and/or consortium of consultants to obtain the best possible product and safely designed structures.

Under the item 'Panel of Experts' it is most essential to have an independent Panel of Experts, with each individual member being an internationally reputed expert in his or her field. There have been problems on some projects where the Panel members were not independent, and were instead hired by the project Consultants. Also, some individual Panel members or the POE as a whole should preferably be involved from the feasibility stage in order to resolve the technical issues early in time (examples: Nyaunggyat Dam - Burma; Kalabagh Dam - Pakistan).

Regarding the item 'Periodic Inspections', it is essential that in addition to the periodic inspections of dam after construction by qualified independent experts, there are periodic and systematic inspections and monitoring by the local professionals responsible for the project, which may necessitate specialized training of the local professionals in the field of safe dam monitoring. Good examples of training of local professionals in dam safety monitoring under the Bank-financed dam project are:

1.   **Nyaunggyat Dam - (Burma)**

2.   **Kulekhani Dam (Nepal)**

3.   **Reservoir Maintenance Facilities Project (Pakistan)**

It has been experienced that dependence wholly on annual inspections by outside experts, without monitoring of the dams project by local professionals, could often result in surprises, e.g., the failure of Bouldin Dam in USA, which failed within about 2 weeks after the Annual Safety Inspection.

Essential Steps for Safe/Successful
Implementation of Dam Projects

Bank experience on the various Bank-financed dam projects indicates that the Bank Guidelines on Dam Safety provide the first essential step and lay down a solid base for safety of dams. There are, however, many other steps which are considered essential for safety on Bank-financed dam projects because of Bank involvement in (i) providing financial and technical assistance to the borrowers, (ii) co-financing packages and contract packaging, (iii) international competitive bidding (ICB), (iv) clearance of technical and contractual clauses, and (v) assisting the borrowers in processing selection of the project consultants, and the Panel of Experts, if requested by the borrowers. These steps pertain to the Bank's much deeper involvement and monitoring of the various stages of the project, including the feasibility/designs/construction. The various steps which are considered essential for safe and successful implementation of Bank-financed dam project could be summarized as:

1. Implementation of Bank Guidelines on Dam Safety

2. Bank staff involvement (with knowledge of dam safety aspects) at various stages of feasibility/designs/construction;

3. Resolution of technical issues before finalization of Designs and Tender Documents.

4. Procurement and contract packaging to suit safe implementation of the project components.

5. Co-financing packaging to suit safe implementation of works.

6. Realistic, effective and successful implementation program including dam instrumentation.

7. Timely resolution of technical issues during construction.

8. Participation in POE meetings and follow-up of the Technical Issues.

9. Continuity of professionals on the project - in the Bank, the borrower, and the project consultants.

10. Training of local professionals for dam safety aspects, instrumentation and monitoring.

Case History

Out of the many examples of successful and safely implemented Bank-financed dam projects, one of the recent examples worth identifying is the Nyaunggyat Dam multipurpose project in Burma. The salient features of the project are:

1. First International Competitive Bidding (ICB) project in Burma;

2. 71 meters high rockfill dam;

3. Multipurpose project for Irrigation, Flood Control, and Power Generation;

4. Financed by the World Bank and four Cofinanciers;

5. Forty-two international civil contractors participated in Pre-bid Conference;

6. Thirteen civil contractors bid under ICB;

7. Seven international contractors worked on project (1 civil, plus 6 E&M);

8.   Project completed 1 year ahead of schedule;

9.   No claims of any kind from any contractor during and/or after the construction of the project.;

10.   After reservoir filling, no technical problems of seepage, cracks, deformations or displacements.

It is unrealistic to expect problem-fee implementation of a dam project. There is no such example in the world. The successful and safe implementation a dam project, however, necessitates timely resolution of technical issues and construction problems before they are created. The achievement of the prevenient schedule and safe implementation of the Nyaunggyat Dam project could primarily be attributed to:

1.   Resolution of technical questions/problems before finalization of tender documents;

2.   Resolution of construction and coordination problems before they were actually created during construction.

3.   Coordinated system of supervision/monitoring of the project. The Project Director was responsible for day-to-day supervision/monitoring of the components, coordinating project consultants, project contractors, and the project staff from Government agencies. The Project Implementation Committee (PIC) provided coordination of the various Government agencies involved in the project (Irrigation Dept., Corporation) and timely resolution of the issues.

Some of the important technical issues which had to be resolved for Nyaunggyat Dam project before finalization of Design and Tender Documents and during the construction of the project included:

Before Finalization of Designs and Tender Documents

1.   Earthfill Dam or Rockfill Dam

2.   Provision or non-provision of grouting gallery

3.   Emergency spillway or fuse plug.

4.   Cofinancing packages and contract packaging for safe and successful implementation of dam project.

5.   Combined or separate tunnels for diversion, penstocks, and irrigation.

During Construction

1. Stability of rock cuts at the intake structure and the powerhouse.

2. Resolution of construction problems in Diversion Tunnel.

3. Protection of spillway toe foundation by stabilized Plunge Pool.

4. Protection of spillway high velocity cavitation on the spillway by provision of aeration galleries.

5. Resolution of provision/non-provision of extended Steel Lining in Diversion Tunnel.

6. Staffing/organization/management/coordination/government agencies /and contractors.

7. Design modifications for Kinda Weir for safety against exit gradient and piping.

If the technical issues are not resolved before finalizing designs and tender documents, it is too late to resolve them during construction with economic alternatives, and it often results in compromising with issues involving dam safety.

Implementation of Dam Safety Measures

From considerations of Dam Safety and the implementation of dam safety measures, the Bank has been involved in different categories and stages of dam projects, including:

1. New dam projects

2. Existing dam projects financed -- not by the Bank.

While on new projects it is easier to deal with enforcing the implementation of Bank Dam Safety Guidelines, the existing dam projects necessitate some additional evaluations and dam safety measures before the Bank could finance the new projects dealing with downstream works, including:

1. Evaluation of the safety of existing dam and the required additional dam safety measures.

2. Emergency action plan for implementation of the identified additional dam safety measures.

Pakistan: **Tarbela Dam:** One of the great examples of the timely addition of dam safety measures and thus ensuring safety of a major dam project is the Tarbela Dam Project in Pakistan, financed by Bank and many Bilaterals. Due to several unusual features of the project, the design and construction of Tarbela was a big leap beyond the known horizons of engineering profession and as such many problems were encountered, some of which were indeed unprecedented:

1. During the very first filling of the reservoir in the year 1974 and the operation of the intake gates of diversion Tunnels 1 & 2, central intake gate of Tunnel 2 got stuck. This ultimately resulted in collapse of tunnel 2 upstream part and consequently serious damage to the outlet chutes and stilling basins of Tunnels 3 & 4. The collapsed portion of Tunnel 2 was rebuilt and basins repaired.

2. The impervious blanket on the bed of the river developed hundreds of sinkholes and cracks during the first filling of reservoir.

3. Intolerable erosion of the hill sides in the spillway plunge pool caused by the high velocity jet (carrying energy equivalent to 40 million horse power at peak discharge) made embayments that outflanked the Service Spillway Flip Bucket.

4. Excessive seepage through the dam foundation and its abutment and the high seepage pressures presented serious problems.

All of these problems were solved successfully by timely action of the Government of Pakistan, the World Bank, and other concerned Bilaterals, using some of the most innovative techniques and implementation of the needed additional dam safety measures as planned and scheduled:

1. A system of sonar surveys was established to detect the formation of any new sinkholes in the bottom of the lake, some 450 feet below the water level. A fleet of bottom dump barges was arranged for dumping soil of suitable specifications to plug any newly developed sinkholes. The system has worked well and all underwater sinkholes have been properly repaired. The number of sinkholes developed was 440 in 1975 which gradually reduced to zero in 1985.

2. The excessive energy carried by the water in Tunnel 4 which was a course of damage to its stilling basin, has been effectively controlled by the construction of modified flip buckets structure (water velocity in the tunnel chute is about 140 ft/ second).

3.   Strengthening of the rock in the plunge pools of both the spillways has been carried out using post-tensioned anchors and massive rollcrete walls approximately 70 feet wide, 200 feet high and totalling more than one mile long. The roller compacted concrete used at Tarbela paved the way for use of this method of rapid construction several new dams the world over.

Other examples of timely and successful implementation of needed dam safety measures on Bank-financed dam projects, thus ensuring the safety of the projects include:

        Curzon dam (India)
        Kulekhani dam (Nepal)

Some of the recent examples of existing dam projects pertaining to categories 2(a) and 2(b) as under **para 12 above**, which have been evaluated for safety of the existing dam projects because of the Bank involvement in financing the downstream irrigation systems and where the required additional safety measures are either under implementation or processing, include:

Cyprus: Evretou Dam: The required dams safety measures are under implementation to ensure safety against the identified problems of immediate concern including: (i) unprotected seepage through outlet drain leading to piping conditions, (ii) severe erosion at the spillway/chute toe endangering stability of the spillway, and (iii) upstream slide at the intake structure.

Turkey: Ivriz Dam: The required dam safety measures are under processing to ensure safety against the identified problems of the Unprotected Seepage through foundations.

Yemen: Marib Dam: The required dam safety measures are under processing for the identified problems of concern: (i) unprotected seepage through foundations, (ii) extremely low drawdown capacity of the outlet, and (iii) stability of the Stilling Basin for full discharges.

Turkey: Kochkopru Dam: The diversion tunnel on Left Bank has collapsed due to piping in pillow-lava and clay layers behind the tunnel concrete lining. While the tunnel is now under construction on Right Bank where the geology is suitable for such structures, the technical issues for safely rehabilitating the project are under review by experts for the needed dam safety measures.

Algeria: Harrezza Dam: The required dam safety measures are under processing for the identified problems of concern: (i) unprotected seepage through foundations leading to piping conditions, and (ii) cracking in the crest of the dam.

Chapter 11

SPECIFIC ACTIONS DURING PLANNING, DESIGN AND CONSTRUCTION FOR DAM SAFETY

William A. Price

There are thousands of tasks and actions required in the process of implementing any project that includes a dam. Almost every one of those actions or decisions made during the planning, design and construction of the dam has some bearing on the overall safety of the dam. Over the past two decades there has been an expanded international awareness of dam safety and more and more examples of how a good dam safety program can work.

Among those many actions, some have a much more significant impact on dam safety and in the following discussion with your today I would like to outline and describe the key activities and actions necessary for a successful dam safety program. Just as important as <u>what</u> should be done, is <u>who</u> should be responsible for performing the actions needed. Bringing a dam project from a concept to full operation is truly a team effort and the manner in which various interests interact may well be the most important element in establishment of a dam safety program that will ultimately save lives, property and economic resources.

The World Bank frequently funds projects with dams and has of course, a major interest in assuring that the dams constructed under their financing, remain in service for the planned life of the project and perform its designed function. A banker's interest obviously centers around making any project a success in order to increase the borrowers capability to repay the loan. There is no question that a dam failure places the borrower in a multiple economic crunch. Firstly the possible loss of life and surely direct property damage as a result of the dam failure. Secondly the loss of the water storage function and reductions or a complete halt of returns from irrigation, power, or municipal water supply and finally there is the additional economic burden of reconstruction or replacement of an in-kind storage facility.

There should be no need to dwell on the desirability of instituting and strengthening dam safety programs with this group, yet we need to keep in mind that a canal, road, power plant or an industrial complex can suffer from lack of maintenance or even partial failure and still not cause the massive and sudden economic disruption that is typical in the case of a dam failure.

The actions and relationships needed for an effective dam safety program are examined here from the perspective of there being a new project with a dam in a borrower country. This is not to imply that all the developed or contributor countries are doing enough or are doing it effectively, but this perspective should allow us to make a comparison of the basics of a good dam safety program to those that many of us have observed in various borrower countries. In this presentation I would like to introduce you to the players, explore their role in the dam safety game and outline some specific actions that are needed to attain an effective program, one which would have an impact on increasing the safety of all dams in the borrower countries.

Dam Safety Laws

The central and vital element of a dam safety program in any country is the legislation that governs the regulatory aspects of dam safety. A program cannot be supported or developed unless it has roots of being a part of the law of the land. The laws cannot cover all the technical aspects of dam safety but they are vital in defining the authority, responsibility and the power of the entity given the lead role.

This primary role or more specifically the enforcement of the law, almost always becomes the responsibility of some designated or created governmental agency. While there have been laws adopted by various countries, it is an unfortunate reality that the initial laws passed often do not contain enough "teeth", which is to say there is not sufficient authority given to the responsible governmental agency and the law remains weak until there has been a prominent dam failure in the country. Following a major dam failure that causes loss of life or results in significant economic damage, then and there comes a public outcry that "there should be a law against letting this type of thing happen again". Although efforts should be made to use examples of dam failures around the world to justify and motivate passage of strong dam safety laws, there are numerous cases in history where the programs have come about only as a reaction to prominent dam failures.

In 1929, it took the St. Francis failure to get the state of California dam safety laws passed and initiate a state-wide program. This program was given strong government support and has provided many contributions over the years in experiences, practices and management of a dam safety program. Yet with strong programs in the states of California, Pennsylvania, Colorado, Idaho and others it still took the Baldwin Hills dam failure to spur national laws in the United States under the 1972 Dam Inspection Act.

Strong U.S. federal support and pressure to strengthen dam safety programs even more occurred following the failure of the U.S. Bureau of Reclamation's Teton Dam in 1976. In the United Kingdom the 1925 laws were in response to failures of the Dolgarrog and Skelmorlie dams. These laws were updated in 1975 in the Reservoirs Act which now provides authority for the regulators to intervene and carry out remedial measures if the owner fails to respond to deficiencies of the inspecting engineers report. In France new regulations were enacted after the Malpasset and Vaiont incidents. Recognizing that the initial filling is a critical period, these French regulations require strict inspections, controlled rates of filling and instrumentation monitoring over the sensitive period when a dam is first placed into operation. India's state of Gujarat experienced the 1979 failure of Machu II Dam due to overtopping in a large flood and suffered the loss of over 2,000 lives. This tragedy has helped create laws and has contributed to a greater awareness in that state and others in India on dam safety.

Involved Entities

When you come into a new ball game, there is the old saying that "you can't tell the players without a program". I believe this is very applicable to the dam safety game. Too often we pick out one of the players and try to blame all the problems on him, whereas it may actually be some other entity that has failed to carry out it's responsibility.

For the purpose of introducing the players, lets assume we have a motivated borrower who has just gone though the process of passing a solid dam safety law. I would like to take you through the process of planning, designing and constructing a dam in country X. Figure 1 lists the entity and outlines the corresponding general action that each is responsible for during the period of implementing a dam project.

Thus these nine entities are my list of key players for dam safety. The entities can take on different personalities from country to country and even within the same country for the various states. Sometimes the players are quite different just because of different projects especially from the owners standpoint. Let us first focus on the governmental organization which by law has been entrusted with the responsibility performing the regulatory function for the purposes of this discussion let us call them the Governmental Dam Safety Organization (GDSO). Obviously the laws cannot cover all the technical elements or procedures, so the GDSO must develop a set of rules and regulations to over the process of dam safety regulation. They will actively participate in adopting design criteria, setting standards, establishment of review and approval procedures as well as determine the notices and enforcement actions to be taken with those who are negligent in following the laws and regulations. The GDSO should be the recipient of all possible assistance by the Bank to encourage its strengthening and effective functioning. The GDSO should also have a strong responsibility after the dam is placed in operation, for monitoring it's condition and safety status for as long as the dam is in operation by establishing and carrying out procedures for regular inspection. The GDSO will therefore be involved in the dam safety actions of the dam from planning through construction and beyond.

Figure 2 outlines the various types of entities that could make up the remaining eight participants. In the U.S. and Europe it is common to see private individuals or private firms in a number of these roles. In developing countries however, it is possible to find situations where almost all of the nine entities are in some form or another a part of the government. Such a situation is really where we need the program to tell the players. There is a need to keep some of the functions and actions of these nine key entities independent. That frequently becomes very difficult when they are all part of the government and particularly when they are within the same department or ministry. However, if the laws are solid and the GDSO is well established, has adequate and well trained staff, who are expert, diligent and honest in discharging their responsibilities, then the dam safety program can still be a success.

Actions Versus Time

The nine entities have varying levels of activities over the project implementation period. All of these actions contribute in some measure to the success or failure of the dam safety program. Figure 3 is a graphical dam safety matrix which illustrates the general importance of specific actions taken by each entity form planning to operation of the dam. The shading of the matrix represents those project phases for which the specific action of the entity is more active and important. A major mistake, or failure of any one the nine entities during these critical stages could lead directly to a dam failure in time, as could a series of shortcomings of all nine of the entities involved. By tracking the specific actions required of these nine entities during each phase of project implementation I believe we can gain an understanding of where some of the more critical links come into being and will help guide us in the Bank in our efforts to strengthen dam safety wherever we are involved.

This overview illustrates the possibility of overlap in activity and to some extent there is. It is the inter-relationship of the actions of these entities that provides "checks and balances" which is the real strength and a positive characteristic of a good dam safety program. The specific actions should be of the nature to assure multiple professional, technical examination of the vital components relating to safety, with as many checks as possible having a degree of independence. This is especially true of the GDSO actions are considered to be important in all phases of the project.

Before outlining the specific actions I believe it would be of interest to examine and keep in mind some of the characteristics of dam failures and near failures of the past as we look at the overall process. The International Committee on Large Dams (ICOLD) has estimated that there are over 150,000 dams in the world that represent a hazard in event of their failure. Since the 12th century A.D. there have been approximately 2000 dam failures, many of course were minor dams while others have caused major loss of life and property. Note that this represents approximately a 1.25 percent failure rate. So far in the 20th century, with larger dams being built, there have been approximately 200 notable dam failures which combined have resulted in loss of life in excess of 8,000 people. Figure 4 shows the relative incidence of failure by dam type for the period of 1990 to 1969 for dams in the U.S. and Western Europe. Note that over this period it looks like a small improvement has been made with the failure rate running slightly over 1 percent. Figure 5 shows the results of an ICOLD survey of dam failures in the world from 1900 to 1975 for dams in excess of 15 meters in height. The failures are broadly classified by cause for the various types of dams. Similar data are presented in figures 6 and 7 that show the distribution of failure or near failure causes for dams in Spain and in the United States.

It is common sense to direct the emphasis on specific actions of a dam safety program toward controlling, upgrading and monitoring factors that influence the past prime causes of failures. It doesn't take a great amount of analysis to see that the major causes of failure are associated with flood hydrology (flows in excess of spillway capacity), geology and treatment of foundations, and elements of the design, selection of material, and operation and maintenance relating to seepage. These three very broad categories have of course many sub-divisions and involve technical specialties that are often extremely site specific yet a successful dam safety program must assure that procedures, policies and methodologies are followed that will concentrate attention in these areas.

Specific Actions During Planning

During the planning phase, dam safety considerations come into play very prominently. The evaluations of design requirements should be made with dam safety principles firmly guiding the process. The public debate, passing of laws, establishment of rules and regulations and cooperative development of technical standards among the engineering and scientific community are continuing activities. For the purpose of this discussion let us assume they have successfully resulted in a strong dam safety law and the establishment of a competent and technically strong GDSO.

The dam failure record points to the need for great effort to be expended on selecting regulations regarding procedures for determining the design flood in relation to the dams hazard potential. while there will be continual refinement in design flood policies, it should be a matter fully settled during the planning phase.

Figure 8 outlines the actions of the entities during the planning phase of a project and gives an indication of how there is multi-agency participation in the flood hydrology component. National and international institutes on meteorology and hydrology as well as world wide practices shape the ultimate policy in the area.

The owners actions int he phase are most critical, as the selection of the engineering team and design review panel are needed to be made and Coordination and cooperation with the GDSO to assure fulfillment of the regulatory requirements will allow the entire project to be implemented more smoothly.

The technical evaluations made in the stage are also very important, especially in effectively and fairly identifying alternatives and selecting the most appropriate dam site.

Specific Actions During Design

During the design phase, the specific actions that impact dam safety become much more direct and obvious. The technical capability of the designers, engineers, geologists and other specialists to conduct a detailed and complete investigation of the site and to apply the latest methods for testing of the foundation and potential construction materials, become extremely important. It is often erroneously thought that this is the only

phase that has a bearing on dam safety. Despite its importance this just is
not the case.

The design team must balance risks against costs and resist owner
pressure if any to minimize project costs at the expense of foregoing the
necessary safety features. The designer must present a sound final design
with clear, concise and accurate contract plans and specifications. Figure 9
lists the actions that should be taken by the related entities during the
design phase. Checks and balances are made in as many areas as possible, and
for that reason the dam review panel and the GDSO participation is vital.
They provide the sounding boards, filters and review in a process that should
openly try to avoid making major errors in design or judgement and should
assure a consistency of safety measures commensurate with the hazard
classification of the dam.

This is the period when there is often the greatest opportunity for the
World Bank as the lending agency to closely monitor the process and to provide
support as strongly as possible to procedures that provide independent
evaluation by the designer, the Dam Review Panel (DRP) and the GDSO. During
the design phase, the construction supervision and O&M units will be
conducting their advance planning in preparation for their more important
future roles.

Specific Actions During Construction

Once a contractor has been selected and mobilized, the center of
attention shifts from the design office to the dam site. The contractor of
course plays no small role in the ultimate safety of a dam. Experience of
successfully building other dams and a record of being responsive to the plans
and specifications is most important in pre-qualifying the construction
companies for bidding on a dam project. The owner needs to select or
establish a strong, experienced construction supervision unit which has the
capability of inspecting, monitoring and motivating the contractor to perform
the quality of work called for in the contract documents.

Figure 10 illustrates the activities of the entities during this period
of time. One of the most critical aspects of this period is to maintain a
continuity of the participation of the designers and the DRP. Key design
personnel must be made available and participate in the critical early
activities of foundation excavation and preparation as well as with final
selection of construction materials. The DPP must also be made available to
review the actions taken during this time.

The GDSO has a general overseeing role and should be given powers to
intervene if the construction supervision activities are grossly ineffective,
even if performed by a sister agency. Note also that the O&M unit should be
in the advanced stages of organizing and preparing for initial operations.

If the establishment of the institutions and the process for selecting
the designer, contractor, and DRP have not been implemented by the borrower

prior to this stage of the project, then the World Bank is faced with some real problems in trying to modify actions that impact dam safety.

Specific Actions in Operations, Maintenance and Surveillance

Although this phase was not included in the title, it is vital for a balanced dam safety program. It is the owner's responsibility to provide the necessary O&M unit with appropriate funding to maintain the dam in peak operation form, to fully monitor its performance and be prepared to respond to any deficiency that develops. Of importance are two specific actions.

1.   maintain in a central location, complete records of the dam's design, construction and operation; and

2.   prepare and periodically update an emergency preparedness plan.

The GDSO becomes the long term watchdog to assure that the dam remains safe. They must have established a program for regular inspections, evaluating performance records provided by the owner and be prepared to order remedial actions which if not performed can result in intervention at the owner's expense.

Summary of Most Critical Actions

The review of the wide variety of actions by the nine main entities during the implementation of a dam, hopefully illustrates that each entity has dam safety responsibilities. In summary, I have selected what I believe to be the 10 most critical of the specific actions. These are listed in Figure 11. If we as members of the World Bank teams involved with projects with dams could assure positive and successful actions in these ten areas, we would surely make a major impact on upgrading dam safety in many borrowing countries.

Figure 1

KEY ENTITIES IN DAM SAFETY

ENTITY	GENERAL ACTION
1. GOVERNMENTAL LEGISLATORS (PUBLIC)	PASS DAM SAFETY LAWS
2. GOVERNMENTAL REGULATORY AGENCY	DRAW UP RULES AND REGULATIONS
3. OWNER OF PROPOSED DAM	PLANS FOR AND SEEKS FUNDING AND APPROVAL OF PROJECT WITH DAM
4. DESIGNER	WORKS FOR OWNER TO INVESTIGATE SITE AND PREPARE DESIGNS
5. DAM REVIEW PANEL	SELECTED BY OWNER TO ASSURE TECHNICAL ADEQUACY OF DESIGNS
6. CONTRACTOR	BUILDS THE DAM
7. CONSTRUCTION SUPERVISOR	OWNER'S REPRESENTATIVE TO ASSURE DAM BUILT TO THE DESIGNS AND SPECIFICATIONS
8. OPERATION AND MAINTENANCE UNIT	OPERATES, MAINTAINS AND MONITORS DAM FOR OWNER
9. LENDING AGENCY.	PREPARES LONG LIST OF COVENANTS AND RELUCTANTLY AGREES TO PROVIDE OWNER WITH 'SWEET LOW INTEREST LOAN'

Figure 2

VARIATIONS OF THE KEY ENTITIES

ENTITY	POSSIBLE TYPE

OWNER	PRIVATE INDIVIDUAL
	COMPANY OR FIRM
	QUASI GOVERNMENTAL
	  ORGANIZATION
	   (CITY,IRRIGATION
	   DISTRICT,COUNTY,STATE)
	CENTRAL GOVERNMENT AGENCY

DESIGNER (ENGINEER OR GEOTECH)	INDIVIDUAL CONSULTANT CONSULTANT FIRM LOCAL OR STATE GOVERNMENT CENTRAL GOVERNMENT

DAM REVIEW PANEL
(BOARD OF
CONSULTANTS
OR PANEL OF
EXPERTS)
	PRIVATE CONSULTANTS
	RETIRED GOVERNMENT
	  ENGINEER
	EXPERT FROM CONSULTANT
	  FIRM

CONTRACTOR	PRIVATE COMPANY GOVERNMENTAL CORPORATION GOVERNMENTAL AGENCY
REGULATORY AGENCY	GOVERNMENTAL AGENCY
LENDING AGENCY	PRIVATE BANK PUBLIC BONDS GOVERNMENT AGENCY REGIONAL DEVELOPMENT BANK WORLD BANK

Figure 3

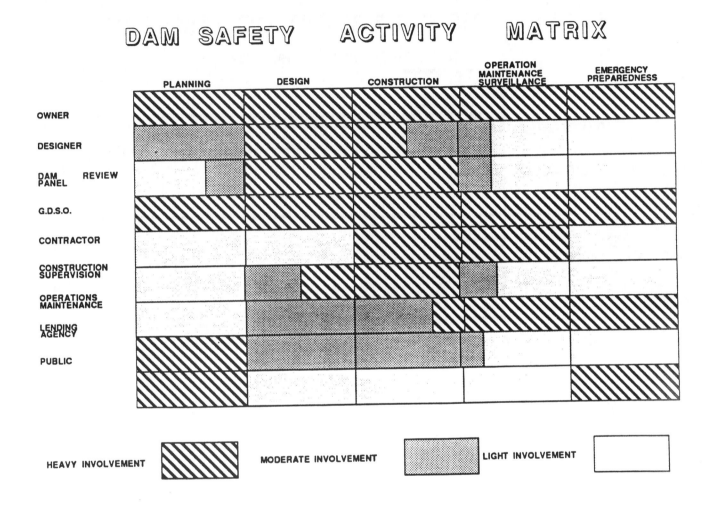

Figure 4

FAILED DAMS IN PERCENT OF DAMS BUILT

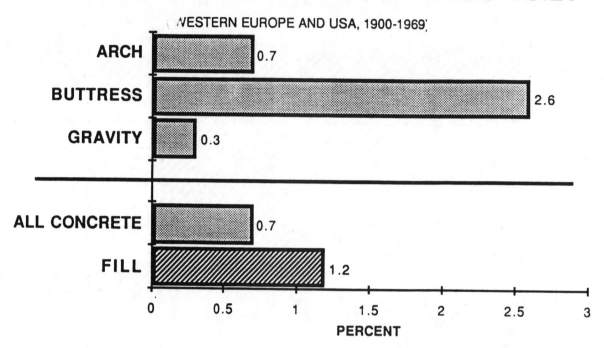

WESTERN EUROPE AND USA, 1900-1969

Figure 5

CAUSES OF DAM FAILURES

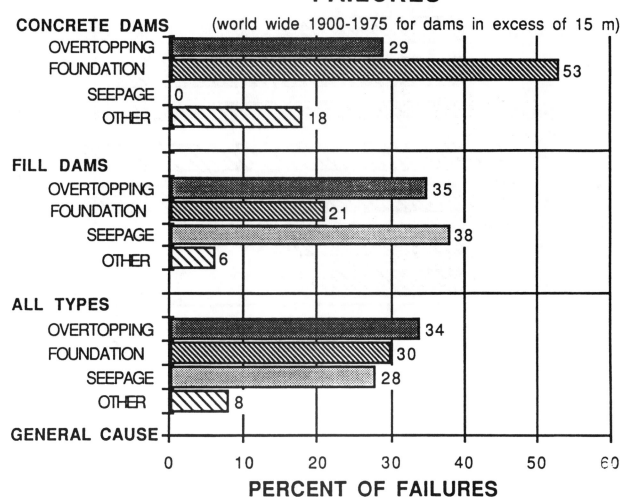

PERCENT OF FAILURES

Figure 6

FAILURES AND SERIOUS INCIDENTS OF U.S. DAMS THRU 1985

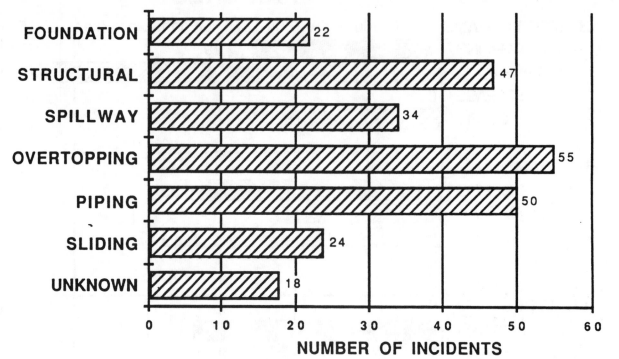

Figure 7

CAUSES OF ACCIDENTS AND FAIILURES OF DAMS IN SPAIN

(1962 SURVEY OF 1,620 DAMS COVERING PERIOD OF 1799 TO 1944)

CAUSE	PERCENT OF FAILURES
FOUNDATION FAILURE	40
INADEQUATE SPILLWAY	23
POOR CONSTRUCTION	12
UNEVEN SETTLEMENT	10
HIGH PORE PRESSURE	5
ACTS OF WAR	3
EMBANKMENT SLIPS	2
DEFECTIVE MATERIALS	2
INCORRECT OPERATION	2
EARTHQUAKES	1

Figure 8

ACTIVITIES AND ACTIONS RELATED TO DAM
SAFETY DURING THE PLANNING PHASE

ENTITY	SPECIFIC ACTIONS
OWNER	JUSTIFY PROJECT SELECT DESIGNER SELECT DESING REVIEW PANEL EVALUATE DAM HAZARD LEVEL NARROW ALTERNATIVES PREPARE ENVIRONMENTAL ASSESSMENT
DESIGNER	FORMULATE ALTERNATIVES CONDUCT FOUNDATION EXPLORATION PERFORM MATERIALS TESTING SELECT DESIGN PARAMETERS PERFORM FLOOD HYDROLOGY PREPARE FEASIBILITY DESIGNS FEASIBILITY COST ESTIMATES PREPARE HAZARD CLASSIFICATION
DAM REVIEW PANEL	REVIEW SITE SELECTION REVIEW DESIGN FLOOD AND ROUTING REVIEW DESIGN CRITERIA
GDSO	DETERMINES REGULATORY RULES AND REGULATIONS SETS PROCEDURES FOR DESIGN APPROVAL ACCORDING TO RISK LEVEL PUBLISHES HYDROLOGIC CRITERIA AND FLOOD DESIGN STANDARDS
CONTRACTOR	MAINTAINS CONTACT WITH OWNER, ENGINEER AND GDSO TO ASSURE HE ACQUIRES SPECIAL CONSTRUCTION TECHNIQUES ASSISTS IN SHARING GENERAL COST DATA
CONSTRUCTION SUPERVISION UNIT	MAKES GENERAL CONSTRUCTABILITY REVIEW
O&M UNIT	SEARCHES FOR STAFF AND TRAINS ESTIMATES OPERATING COST PLANS OPERATIONS CENTER, HOUSING AND EQUIPMENT
LENDING AGENCY	ASSURES THAT OWNER HAS PLANNED FOR COST OF DAM, DRP, DESIGN, O&M AND IS EVALUATING CRITICAL ISSUES

Figure 9

ACTIVITIES AND ACTIONS RELATED TO DAM SAFETY DURING THE DESIGN PHASE

ENTITY	SPECIFIC ACTIONS
OWNER	ISSUES DIRECTIVE TO DESIGNER OF PHILOSOPHY ON RISKS AND HAZARDS TIES DOWN FINANCIAL ARRANGEMENTS PREQUALIFY BIDDERS
DESIGNERS	FINAL SITE EXPLORATION AND TESTING MATERIALS EVALUATIONS SITE SELECTED TYPE OF DAM SELECTED FOUNDATION DESIGNS PREPARED MAIN SECTION DESIGNS (STABILITY, SEISMIC) OUTLET WORKS SIZING PREPARE FINAL HAZARD/RISK ANALYSIS FLOOD HYDROLOGY AND SPILLWAY DESIGN PREPARE CONTRACT DOCUMENTS(PLANS AND SPECIFICATIONS) REQUEST APPROVAL FROM GDSO PREPARE O&M MANUALS COORDINATE WITH CONSTR. SUPERVISION UNIT ASSIST IN TENDERING AND BID EVALUATION
DAM REVIEW PANEL	REVIEW DESIGN CRITERIA, PARAMETERS AND METHODOLOGIES REVIEW FOUNDATIONS DESIGNS REVIEW MATERIAL SOURCE SELECTION REVIEW MAIN SECTION DESIGN REVIEW FLOOD HYDROLOGY/SPILLWAY DESIGN ADDRESS PROBLEM AREAS AS REQUESTED BY OWNER OR ENGINEER
GDSO	MAKES AVAILABLE RULES AND REGULATIONS MEETS WITH ENGINEER TO CLARIFY TECHNICAL REQUIREMENTS REVIEWS AND APPROVES FLOOD HYDROLOGY AND SPILLWAY DESIGN REVIEWS GEOTECHNICAL ELEMENTS OF DESIGN REVIEWS PLANS AND SPECIFICATIONS AND GIVES APPROVAL TO START CONSTRUCTION

CONTRACTOR	TRACKS DESIGN STATUS AND EXAMINES ANY SPECIAL CONSTRUCTION TECHNIQUES NEEDED PREQULAIFIES AND PREPARES BIDS
CONSTRUCTION SUPERVISION UNIT	ASSISTS IN PREPARING CONTRACT DOCUMENTS PREPARES CONTRACTORS AND OWNERS AREAS AT THE CONSTRUCTION SITE ASSISTS IN BID EVALUATION
O&M UNIT	CONTINUES TO PLAN AND PREPARE FOR OPERATION ESPECIALLY WITH INITIAL FILLING PROCEDURES PREPARES OPERATIONS AND MAINTENANCE PLAN ASSURES BUDGET AND STAFFING FOR O&M
LENDING AGENCY	ADVANCES PROJECT APPRAISAL AND EVALUATION OF CONDITIONALITIES OF LOAN COMMUNICATES DAM SAFETY CONCERNS TO THE BORROWER(OWNER)

Figure 10

ACTIVITIES AND ACTIONS RELATED TO DAM SAFETY DURING THE CONSTRUCTION PHASE

ENTITY	SPECIFIC ACTIONS
OWNER	ACCEPTS OR REJECTS BIDS ARRANGES FOR NECESSARY PERMITS MOBILIZES CONSTRUCTION SUPERVISION TEAM FINALIZES FINANCIAL ARRANGEMENTS
DESIGNER	PROVIDES TECHNICAL SUPPORT TO CONSTRUCTION SUPERVISOR UNIT COMPLETES NECESSARY SUPPLEMENTAL DRAWINGS ASSISTS IN INTERPRETATION OF FIELD TESTS DIRECTLY PARTICIPATES IN FOUNDATION APPROVAL RESPONDS TO CHANGED FIELD CONDITIONS WITH MODIFIED DESIGNS REFINES O&M MANUALS EVALUATES AND APPROVES ANY CONTRACTOR ALTERNATIVE PROPOSALS TAKES LEAD IN DEVISING EMERGENCY PREPAREDNESS PLAN
DAM REVIEW PANEL	REVIEWS FIELD CONDITIONS UPON OPENING OF FOUNDATION REVIEW THE NEED FOR DESIGN REVISIONS EVALUATE CONSTRUCTION QUALITY RESPOND TO PROBLEM AREAS IDENTIFIED BY OWNER, DESIGNER, OR CONSTRUCTION SUPERVISOR
CONTRACTOR	CONSTRUCTS DAM ACCORDING TO PLANS AND SPECIFICATIONS REPORTS ANY APPARENT UNSAFE CONDITION AND REQUESTS CLARIFICATION ON MATTERS THAT MIGHT HAVE BEARING ON DAM SAFETY
CONSTRUCTION SUPERVISION UNIT	ASSURES QUALITY OF CONSTRUCTION PERFORMS DETAILED INSPECTION OF FOUNDATION AND MATERIAL SOURCES MAINTAINS ACCURATE AND COMPLETE RECORD OF CONSTRUCTION ACTIVITY AND TESTS (INCLUDES PHOTOGRAPHIC RECORD) COORDINATES WITH DESIGNER ON MATTERS INVOLVING DAM SAFETY ANALYSIS KEEPS GDSO INFORMED OF STATUS

GDSO	MAKES SITE INSPECTIONS AT KEY TIMES ASSURES THAT ADEQUATE CONSTRUCTION SUPERVISION IS BEING APPLIED PERFORMS COMPLETION INSPECTION PRIOR TO ALLOWING INITIAL FILLING REVIEWS AND APPROVES ALL MAJOR DESIGN MODIFICATIONS
O&M UNIT	OBSERVE CONSTRUCTION INSTALL OPERATIONS HEADQUARTERS AND ACQUIRE EQUIPMENT FOR OPERATIONS ASSIST IN ACCEPTANCE TESTS ASSEMBLE O&M MANUALS AND AS-BUILT DWG PREPARE OPERATIONS AND MONITORING RECORDS SYSTEM ESTABLISH OPERATING RULES PARTICIPATE IN FINAL INSPECTION AND INITIAL FILLING
LENDING AGENCY	SUPERVISE OWNERS ACTIONS AND PROJECT PROGRESS AND QUALITY ASSURE THAT DRP IS ACTIVE AND EFFECTIVE CALL IN SPECIAL CONSULTANTS TO EVALUATE SPECIAL PROBLEM AREAS
PUBLIC	SHOULD BE GIVEN ACCESS TO OBSERVATION AREA DURING CONSTRUCTION SHOULD BE PROVIDED INFORMATION ON PROJECT FUNCTION PARTICIPATE IN PUBLIC HEARINGS OR DEBATE ON DAM SAFETY CONCERNS PARTICIPATE IN DEVELOPMENT OF EMERGENCY PREPAREDNESS PLAN

Figure 11

SUMMARY OF TOP TEN ACTIONS REQUIRED FOR EFFECTIVE DAM SAFETY PROGRAM

1. PASS FIRM DAM SAFETY LAWS WITH PROVISIONS TO FACILITATE ENFORCEMENT AND TO HOLD OWNER ACCOUNTABLE.

2. ESTABLISH WELL TRAINED, FULLY STAFFED AND ADEQUATELY FUNDED GOVERNMENT DAM SAFETY ORGANIZATION (GDSO) WHICH IS <u>INDEPENDENT</u> OF DESIGN, CONSTRUCTION OR OPERATING ENTITIES.

3. REQUIRE A DAM REVIEW PANEL WHOSE MEMBERS HAVE RENOWN EXPERIENCE IN VARIOUS CRITICAL ASPECTS OF DAMS SIMILAR TO THE DAM BEING DESIGNED AND CONSTRUCTED.

4. ASSURE THAT LEAD DESIGNERS, GEOLOGISTS AND MATERIALS SPECIALISTS ARE AVAILABLE AND INVOLVED ON SITE DURING KEY DECISION PERIODS UNDER THE CONSTRUCTION PHASE.

5. ASSURE THAT THE GDSO HAS AN ACTIVE AND <u>INDEPENDENT</u> ROLE STARTING FROM ESTABLISHMENT OF RULES AND REGULATIONS TO THE LONG TERM MONITORING OF THE DAM'S PERFORMANCE.

6. ESTABLISH PROCEDURES WHERE CHECKS ARE MADE BY THE DESIGN AND CONSTRUCTION SUPERVISION PERSONNEL TO MINIMIZE THE POSSIBILITY OF TECHNICAL ERROR.

7. PROVIDE AN ATMOSPHERE FOR TECHNICAL TRAINING, TECHNOLOGY TRANSFER AND ADVANCEMENT OF PERSONNEL CAPABILITIES IN DESIGN,CONSTRUCTION SUPERVISION AND OPERATION AND MAINTENANCE OF DAMS ON A COUNTRY-WIDE BASIS.

8. ASSURE THAT CONSTRUCTION QUALITY IS GIVEN EXTREMELY HIGH PRIORITY AND ATTENTION AND THAT AN ADEQUATE O&M PLAN IS DEVELOPED EARLY AND FOLLOWED.

9. ASSURE THAT COMPLETE RECORDS ARE ASSEMBLED AND MAINTAINED BY THE OWNER TO INCLUDE SITE INVESTIGATION DATA, DESIGN DOCUMENTATION, CONSTRUCTION RECORDS, TEST RESULTS, AS-BUILT DRAWINGS, CONTRACT SPECIFICATIONS, CHANGE ORDERS, INSTRUMENTATION MEASUREMENTS AND O&M RECORDS. SELECTED RECORDS SHOULD BE FURNISHED TO AND MAINTAINED BY THE GDSO. SELECTED RECORDS SHOULD ALSO BE KEPT ON FILE AT THE DAM.

10. ASSURE THAT AN EMERGENCY PREPAREDNESS PLAN IS PRODUCED IN COOPERATION WITH LOCAL OFFICIALS AND THAT IT IS REGULARLY REVIEWED AND UPDATED. EMERGENCY OPERATING PROCEDURES SHOULD BE MAINTAINED AT THE DAM.

Chapter 12

RESERVOIR INDUCED SEISMICITY: A DAM ENVIRONMENT SAFETY ISSUE

Thomas Vladut

Reservoir Induced Seismicity (RIC) was considered as a strange environmental phenomenon in relation to new impoundments which confounded the scientific community for years. Today it is recognized that RIS is an environmental issue which is largely understood and therefore consequently should be used in the design of dams. The control and mitigation of environmental interference is related to the social and technical awareness of the designers who are and will be involved in the design of new impoundments.

Environmental risks are often evaluated by the real or discerned damages and hazard to nature, health, life threats and potential economic losses. Environmental interferences of RIS are associated with the concept of dam safety and the cost implications of this are difficult to estimate for specific environmental issues. The triggering of induced seismicity associated with large dams has rarely jeopardized the safety of such structures, notwithstanding the destruction of Koyna Dam in 1967 with the loss of some 200 lives.

The unintentional triggering of small scale activities is a common characteristic of reservoirs and other engineering developments, such as petroleum reservoirs, underground mining extraction, waste disposal and fluid withdrawal. Related to other seismic activities the phenomena is known also as Man-Made Associated Seismicity (MMAS). The first underground operation to be aborted on environmental grounds due to seismic activities was a deep well for disposal of contaminated fluids at Rocky Mountains Arsenal (Denver Colorado).

RIS compared to other environmental concerns is characterized by a direct relation between cause and effect with a short time span to identify the interference. Frequently traditional environmental concerns dealing with predominant long term effects, as health deterioration, pollution etc. are evaluated statistically. The statistical interpretation makes it possible in some instances to be perceived as threats, while through traditional engineering procedures such risks could be avoided, e.g. the "predicted environmental disasters" caused by arctic pipelines.

RIS frequently gives faster indications about the generating causes in new impoundments, new disposal or intensive withdrawal.

This paper will focus on the implications of RIS related to dam safety and associated risks relating more to the economical implications rather than detailed design and risk mitigation procedures.

Mitigation is the object of technological procedures to accommodate a specific risk like RIS and are often implemented by engineering code-practice. The engineering codes in comparison with economic parameters take longer to mitigate specific environmental concerns.

The development of RIS identification-mitigation procedure is an example of environmental awareness and successful approach to environmental protection for reservoirs. The experience in controlling such an environmental hazard is an effective datum for professionals in their ability to curtain similar environmental risks which are inherent with any significant development.

The awareness of RIS control facilitates the protection of the environment. This preventive measures considered systematically and addressed along with the environmental management programs of the Bank or during Environmental Impact Assessment procedures, will contribute to the reduction of environmental degradation and will achieve environmental enhancement in addition to economic benefits of specific projects.

Most environmental concerns in relation to global climatic changes, deforestation, watershed degradation, salinity, decertification etc. are directly or indirectly connected with water resource development and management practice in which reservoir utilization is a significant components. The concern of mitigation of RIS hazards could be one of many components in integrated environmental management by which international organizations (like the World Bank) may contribute to the preservation and development of the environmental enhancement process.

Background of Development regarding Environmental Concerns Associated with Dams and Their Reservoirs

Occurrence of RIS is related to the development of new reservoirs mainly to provide water and hydropower. The incidence of unintentional triggering of seismic activities could be considered through the many unforseen inherent nuisances of an environmental nature. It took several years before the interference was recognized and technical understanding and awareness was developed.

In recent times, dam and reservoir engineering experienced modifications *vis-a-vis* the changes in economic and social requirements. In the first instance the economic nature was associated with increased demand for power and water. This was followed by important challenges in a more environmentally conscious and sensitive society.

Environmental social conscience with associated demands brought about supplementary components which affected the cost when impacts were unavoidable and the trends focussed on indicated potential hazardous conditions.

The occurrence of RIS is an issue associated to dam and reservoir design. Dam design involves many uncertainties, albeit the majority of dams are safe structures. The uncertainties in engineering arises from the inherent variability of natural phenomena. The target of dam design and safety programs is to provide structures which have an acceptable probability of serviceable utility and safety throughout their life. Design needs to encompass simultaneously, aspects of operation, environmental modifications and maintenance, based on a concept of security, which could not avoid the possibility of failure.

Properly designed, constructed and operated dams ensure the safety of downstream developments in the event of flooding. This increased safety is achieved by creation of a new retaining structure which at the same time replaces the real associated with the hazard of eventual dam failure.

Unfortunately like many other human enterprises, dam structures are not risks free. The dam design process became and undertaking consistently dealing with a number of uncertainties. Most of uncertainties are determined by statistical trend evaluations.

When design is associated with unknown conditions, dam safety evaluation follows a similar probabilistic concept, either in the form of safety coefficient or in the expression of failure risks. Such probabilistic concepts of structural failure are best exemplified by Environmental Impact Assessment (EIA) procedures by which hazard definition usually includes "dambreak analysis" and inundation hazard maps.

Dam safety analysis relate to environmental parameters such as: Inflow Design Flood (IDF), Probable Maximum Precipitation (PMP), storm development conditions, Probable Maximum Design Earthquake (PMDE), etc.

For specific environmental conditions associated with dams and reservoirs, two different stages become significant.

Modification of the Reservoir/ Structural Environment by New Development

It should be noted that this modification process is of a dynamic nature and can be painful. The construction phase is notably damaging to the local surrounding areas and takes an even more dramatic development with the initiation of impoundment and the shock to the ecological habitat.

Note on the optimistic side, good preservation procedures take advantage of natural strength in reaching a new equilibrium. Estimation on the time to reach a new equilibrium condition is difficult to quantify mainly because of the multitude of interrelated components.

The New Environmental Equilibrium and Long-Term Interference
Developed Throughout the Structural Aging Process.

The process of Environmental Impact Assessment (EIA) has been developed in response to increased public concerns regarding the environment. EIA provides the means of hazard analysis of certain probable risks during the licensing and financing process of major industrial developments. Awareness and risk identification are related to the evaluation of potential mitigation procedures for specific identifiable hazards.

Dam design concerns focussed on the object of enhancement of environmental conditions with a view to the preservation of initial conditions with their correspondent procedures of structural ageing, long term environmental implications start to be of significant concern for dam safety: water quality, reservoir ecology with specifics on fisheries and wildlife habitat, downstream effects and new usages for recreation. All these complex issues make it necessary to reevaluate the economics of the real cost of environmental protection and better quantification of the benefit to the environment.

This benefit-risk, associated with environmental protection and enhancement, is increasingly desirable for consideration during the EIA process in an economical format for cost-benefit analysis.

Background on Reservoir Induced Seismicity

Awareness and Understanding

Reservoir induced seismicity (RIS) is usually associated with large reservoirs, but has also been observed at a number of smaller water storages with lower heads of 60 to 40 m and less. Historically, the development of large reservoirs for hydro plants was the first type of engineering activity which significantly affected the earth's crust and RIS is associated with such stress modifications.

The first reference related to the risks of triggering seismic occurrences in new storages is associated with the Quedd Fodda Dam in Algeria (1932, H height of dam = 101 m., V reservoir volume = 0.22 Kmc., M magnitude = 3, 1 year after Impoundment), where the instrumentation estimated the location of the seismic source as very superficial at 300 m. Other delayed seismic events were related to the impoundment of the Marathon Dam in Greece, 1929, (H = 76 m., V = 0.04 Kmc., M = 5.7), significant shocks were recorded nine years after the reservoir impoundment.

Carder, D.S. (1945), was the first to point out the association of seismic activities with water storage elevation variation and loading for the Hoover Dam's reservoir (Lake Mead) where seismicity started three years after impoundment (1935, H = 221 m., V = 36.7 Kmc., M = 5) with further occurrences of earthquakes of less intensity over a period of years.

During the 1960s several cases of RIS were detected some with
significant seismic effect.   Most notably were:

```
Hsinfengkian - China 1959            H = 105 m   V =  13.8 Kmc   M = 6.0 after 3 years
Kariba      - Zambia/Zimbabwe 1959   H = 128 m   V = 175.0 Kmc   M = 6.0 after 5 years
Kremasta    - Greece 1965            H = 160 m   V =   4.7 Kmc   M = 6.3 after 3 years
```

In the early part of the 1960s particular notice should be taken in
relation to the maturing of a new branch of geological sciences: rock
mechanics. A number of rock mechanical advances were directly associated to
unfortunate events in dam engineering, in particular with the failure of the
Malpasse arch Dam (1960, 400 lives lost) and the overtopping of the Vajont
arch dam due to the slide of Mont Toc into the reservoir with significant
losses in human lives (about 2000).

Significant attention was associated with the Denver Colorado earthquake
which was considered to be triggered by the injection of waste fluid in deep
disposal wells (1962 - 1966, Depth 3671 m., V = 0.5 Kmc., 3 events M = 5.2 and
1584 events M = 1.5 - 4.4). Disposal was fluid injection and earthquakes.

The relation was pointed out by Evans (1965), in the form of a
correlation between the monthly tremor frequency and the monthly amount of
disposed water by injections for a period of over three years. The frequency
of earthquakes, diminished after injection ceased, however, shocks continued
up to three years later. Not all seismic activity was proved to be connected
with the fluid disposal but awareness and social sensibility brought the
operation to an end due to environmental consideration.

Self-generated earthquakes by Koyna Dam's reservoir, India 1962,
(H = 103 m., V = 4.75 Kmc., M = 6.3 after 3 years) damaged the gravity
structure (claimed 200 lives, injured over 1,500, 80% houses destroyed and
unhabitable) and caused damaged in the suburbs of populated areas as far as
230 km from the epicenter.

Increased perception of potential social risks was raised by Rothe's
article, 1968, "Fill A Lake, Start An Earthquake". In his paper he associated
RIS potential to reservoirs with significant water depth over 100 m,
development of seismic activity months/years after impoundment and favorable
geological conditions. The geological hazard was associated with the presence
of brittle rocks and the presence of activable faults.

As understanding and technology developed, not surprising was the
realization of the first controlled earthquake experiment (1967 - 1970)
carried out in the Rangely oil fields (Colorado) where earthquakes (M = 0.5
and 3.5) occurred within the injection zones used for petroleum secondary
recovery.

Knowledge Based on Increased Number of Case Histories

Significant scientific activities developed on a global scale with discussions generated during and after specialty reunions dedicated to RIS (UNESCO, 1970 "Working Group on Seismic Phenomena Associated with Large Reservoirs"; Royal Society London, UK 1973 "International Colloquium on Seismic Effects of Reservoir Impounding"; 1975 Banff, Canada "The First International Symposium of Induced Seismicity"). Important contributions provided milestones of understanding and demonstrated the role of disequilibrium of pre-existing natural stresses associated to reservoirs, by the part played by the weight of the stored water (Gough & Gough 1970) and increase of pore fluid pressure (Raleigh 1972 & 1976) in triggering seismic activities. The RIS literature on technical aspects and case histories has over a thousand pertinent references, but studies could not be attempted without the comprehensive book dedicated to RIS **"Dams and Earthquakes"** by Gupta H.K. and Rastogi B.K. (1976, p. 229 Elsevier).

As more cases of RIS were identified more general observations became possible. To note a study in association to seismic risks for the Auburn Dam project, Woodward-Clyde Consultants evaluated 57 case histories and classified them in relation to the pertinence of data supporting the RIS occurrence (accepted and related either to microearthquake or macroearthquake events, questionable and some not related). Such classification of RIS became more specialized on the size of the earthquake and to the change of microearthquake activity (six cases, 1982 Gupta, H.K.) or related to other causes, fluid injection and withdrawal.

Accumulation of case histories made it possible to study the phenomena on a more general basis, statistically using water depth of reservoirs and volume of impoundment, e.g. USGS Open File report 1976, which included about 1800 reservoirs and deciphering relationships between frequency of RIS and depth of reservoirs (e.g., 26% for storages between 150 and 200 m).

Now in the 1980's there is a good understanding of elements controlling RIS, but there are still some difficulties in specific cases of application for dam design and utilization for dam safety needs (the Manicougan five reservoirs were predicted as hazardous for induced seismic risk without developing associated seismic activity).

The World Congress of ICOLD (International Committee on Large Dams) New Delhi 1979, discussing the problem of seismic risks, supported the undertaking of a world wide survey on reservoirs affected by RIS. The survey produced specific elements, mostly on the geomorphological conditions of reservoirs in an environment in which several new geotechnical methods indicated better understanding and predictability, using a improved observational method of analysis of ground related structure-ground interaction (Geotechnical Observational Method, R.Peck 1976 Rankine Lecture).

Improved procedures for understanding ground responses associated with stress modifications (stress path method) became closer to practical utilizations. Several in-depth researches of case studies e.g., Schlegais,

Austria and stress evaluation involving reservoirs up to 12 km in depth, at reservoir emptying and refilling with greater stresses than any previous time and position, (Withers, 1977) provided supplementary knowledge.

Full utilization of several interrelated elements required the enhancement of the comparative means for reservoirs of different shapes and sizes and diversity of geological conditions. Simplified description of valley forms were associated to initial procedures of arch dam design (Tolke) and were developed for volume, size comparison in a non-dimensional approach.

In the desire for generalization, the new impoundment was hydraulicly treated as a new underground source of infiltration. The pore water pressure development is described as a follow up to the unsteady nature of the infiltration. Technical details on the specific elements are detailed in Q60 R40 which presents the Risk Prediction Model for RIS.

Application of RIS Risks Evaluation

Any kind of engineering procedures would have questionable values if proven successful verifications are not available.

The pattern for RIS development is perceived as a case of massive hydrofracturing due to impoundment an some time due to water storage operations. The understanding provided by the Risk Prediction Model (RPM described in Q60 R40) are derived from a relatively good correlation of field data on the evaluated depth of the seismic sources of RIS nature.

The Risk Prediction Model approach is based on three aspects of reservoir impoundment:

1. The impoundment of new reservoir is considered as a new underground source of flow and the flow conditions and pore water pressure development are derived from the results of the unsteady confined source development;

2. The size of reservoir is simplified to facilitate morphological comparison of different shapes of storages. It could be noted also that during the impoundment of reservoirs, important shape changes take place as storage is filled from the lower watershed to the upper part of the valley;

3. The evaluation of the depth of underground storages affected by the weight of the new reservoir and the pore pressure development is of geomechanical nature and allows geomorphological comparison of reservoir conditions. For six case histories a close relationship between the field data and depth estimated by model was obtained (Kremasta, Hoover, H. Verwoerd, Koyna, Grandval and Schleighies).

To estimate the geological particularities for assessment of the deformation below the storage generated by the weight of water, instead of traditional elasticity modulus or deformation, the coefficient of Poisson was used. This coefficient relates better to stress conditions in circular self-reinforcing environments and by this modeling the stress development in the crust and the brittle nature of the ground. Certainly difficulties are related to estimations of these parameters and will also be connected to the geological structure of the site.

The ICOLD survey provided details on geomorphological conditions of reservoirs accepted as pertinent to RIS development. Most RIS affected reservoirs (twenty) are concentrated in a specific domain of a nondimensional comparative diagram. The area of incidence for reservoirs affected by seismic activities was considered and area of risk for new reservoirs with similar geomorphlolgical conditions. This identification of risk of RIS incidence has a statistical nature (Criteria A) and must be further associated to other risk factors related to the effective depth and size (volume) of the new water storages (Criteria B and C).

One of the first applications of the Risk Prediction Model was used for the reservoirs on the Manicougan River (Quebes M3 and M5 which exhibited different seismic behaviors. For the Manic 5 (H = 214 m, V = 35.7 Kmc) reservoir, one of the world largest reservoir, induced seismicity was expected but the impoundment produced no seismic events. The much smaller reservoir Manic 3 (H = 108 m, V = 0.35 kmc.) situated in a similar geological setting, with 100 times less volume, seismic activity of magnitude 3-4 (M = 4.1 at impounding) was detected. The smaller M3 reservoir is located within the risk of RIS area and the larger reservoir M5 is outside the hazard incidence zone. This explanation of different behavior of RIS was a post impoundment identification which endorsed the applicability of the Risk Prediction Model. During the post impoundment RIS identification for Manicougan, a further step was taken to assess continued behavior for new reservoirs in particular LG2, LG3 and LG4 on the La Grando River, James Bay Development, Quebec, Canada.

The first confirmation on the predicted seismic behavior was obtained for the LG2 reservoir which allowed detailed analysis of the model potential. The impoundment implies a continuous sequence of modification of the shape of the storage. for each state of impoundment, in particular for each increase in depth, a different stress condition applies. The process of reservoir impoundment will correspond to a continuous stress modification and an associated access path could be identified on the risk identification diagram. Even if the final stage may be in the risk area, several intermediate situations may not be risk prone. A good similarity between the potential hazard and seismic events was obtained, reinforcing the definition of the limits of the potential RIS risk prone areas. Even so, risk limits should be considered as of a statistical nature. The limits of the risk prone area should be verified further on other RIS reservoirs and on new storages. Such supplementary data would improve the statistical knowledge and enhance further applications.

Potential of Environmental Mitigation, Control of RIS and the Role of Environmental Impact Assessment Process

In the assessment of seismic risk, some mitigation procedures can be implemented. The process of RIS evaluation is controlled by two activities:

1. Hazard identification with risk assessment for potential seismic activities;

2. Development of passive mitigation procedures.

Hazard Identification

The process is directly related to regulatory procedures of the Environmental impact Assessment (EIA) aiming at identifying the RIS hazard and associated risk components (Criteria A,B,C) for the purpose of considering mitigation procedures required by the design and development process.

The risk of man-made seismicity is important mainly in areas with low natural seismicity. The highest reservoir induced seismic activity reached about six on the Richter scale. Other types of underground reservoir operations associated to petroleum recovery have reached a level of 3, (reference to Rangely oil field operation), 4 for mining deep underground operations and 3.3 for fluid withdrawal (maximum 3.3 in Goldsmith, Texas).

Mining related induced seismicity received more attention as mines became deeper. Two international symposiums (1984 Johannesburg, 1988 Minnesota) provide accumulated experience on this specific concern.

RIS hazard identification is related to supplementary economic requirements in providing added seismic protection for the dam structures. Often this increased protection could be in the range of several percent of the total structural cost. Implementation of such supplementary commitments required by the seismic design to provide added safety conditions, should be site specific. This requires the identification of hazards related to the natural hazard of seismicity (N) and the potential risks of induced seismicity (M) by means of a simple comparative criteria, ratio R, $R = N/M$. Both elements are expressed in earthquake magnitude e.g. on the Richter scale. One would refer to the natural hazard of earthquakes (N) which could be evaluated in a preliminary manner from the usual seismic risk maps or similar values generated by specific seismologic and specialty geologic evaluations. Evaluating the man-made seismic risks (M) the simplest procedure would be to use the known historic maximum value of triggered RIS events, 6 or 6.4 on the Richter scale for water storages or the correspondent maximum values for activities involving mining, water withdrawal and petroleum related recovery processes.

For areas with greater risk of natural seismicity, man-made seismicity may only produce the change of the pattern of frequency of the natural seismic activity. Such changes may occasionally be of a beneficial nature, seismic gaps generated by some new reservoirs being acknowledged for some dams:

Anderson   - USA   H = 72 m   V = 0.10 Kmc
Flaming Gorge   - USA   H = 153 m   V = 4.67 Kmc
Mangala & Targela (Pakistan, Gleen Canyon) - USA   H = 216 m   V = 33.30 Kmc

Development of Passive Mitigation Procedures

Increased awareness of the RIS risks has led to some passive mitigation procedures using slower rates of filling during the last decade. The practice is also associated with the construction of earthfill dams, where a slower impoundment is required to avoid excessive pore water pressure accumulation in the earthfill structure.

More structured mitigation procedures are associated with the control of the rate of impoundment, or for more risk prone conditions, a design of the rate may be implemented. Such design procedures relate to the stress path method as used for the evaluation of seismic activity occurrence for the LG2 reservoir.

Up to now the most important preventive components are related to impoundment and reservoir operation monitoring of seismic activities. This brought about understanding of specific site conditions and contributed indirectly to some tempering of filling rates for some reservoirs.

The actual risk evaluation refers to the frame of Environmental Impact Assessment procedures and in the future it will be difficult to allow for the unintentional nature of man-made seismicity.

Without providing a fully risk-free approach, the frame of hazard reduction procedure could require the following steps:

1. Definition of risk associated to the natural seismic background of the sites with risks smaller or larger than potential events generated by an RIS activity. The estimation of risk for the principal types of development prone to RIS (reservoirs, petrol, mining etc.) would be through historically known limits. Even if it is difficult to accept higher limits that recorded in the past, it should be noted that no direct relationship between seismic magnitude and reservoir sizes are available outside of statistical correlations and therefore some uncertainty will remain. Some theoretical studies to estimate the upper limits of RIS events similar to other dams safety concerns like the evaluation of maximum flood values is necessary.

    The analysis will generate two different approaches:

    (a) For natural seismic areas risks higher than in the past generated by RIS, no direct supplementary cost may be expected. Some consideration should be given tot he

increased frequency of seismic activity in an intermediate-
max. to the maximum design earthquake range which may affect
long term structural behavior. This potential may shorten
the life of the structure or may determine the requirement
for rehabilitation or increase the cost of maintenance.

(b) The second option for zones of low natural seismicity is
associated with the necessity to detail the risk for such
potential RIS hazardous situations.

2. Definition of RIS risks

3. Evaluation of the risk of RIS is associated with three distinct
criteria of quasi statistical nature:

(a) Identification of potential incidence of RIS on the
geomorphological structure in comparison with similar
reservoirs affected by seismic activity associated to
impoundment (Criterion A), this may provide a first
indication of the potential risk. The design estimate on
the size and probability of RIS will be made on the grounds
of two supplementary criteria related to the size and depth
of the new reservoir.

(b) Two statistical (Criteria B and C) refers to the new risk
elements for smaller reservoirs affected by RIS.

Mentioning smaller reservoirs reference is made to the old rule of thumb
for RIS incidence associated with dams with more than 100 m high retention and
reservoir greater than one kilometer cub (kmc).

The existence of risks associated with lower head reservoirs was known
form the early 1960s when several shallow reservoirs were associated with RIS
e.g. Marathon, 67m/M=5.7; bajina Basta 90/M=4.5-5; Camarillas 49m/M-4.1; Clark
Hill 60m/M=4.3; Kerr 60m/M4.9; Grandipet 36m/M=3.50.

Also some shallower reservoirs were identified with depths of only 30m
(Q51 R36 Garcia Yague, 1979). Similar evidence is accumulated for reservoirs
with volumes less than one kilometer cube of water.

Examples of such component risks evaluation using all criteria (A,B,C)
are available for a typical continental zone of low natural seismic risk,
(Alberta, Canada in Q60 R40) and are related to 26 future reservoirs.

The Probabilistic Nature of RIS Identification

Hazards Associated to RIS

The implications of RIS risks are linked to the complex nature of
providing safety margins for dam structures. The safety factors related

requirements will be of economical significance with new structures in zones with low natural seismicity. General risks of dam failure are often expressed in a probabilistic manner. Failure frequency is higher during the early years of operation and the rate of dam failures during the first filling of reservoirs has been adjudged at 1:300.

Risks of RIS incidence have been expressed in probabilistic terms for reservoirs with heads of more than 100 m. This probabilistic evaluation is related to the potential size of earthquakes triggered in comparison with the size of reservoirs. This statistical approach will be associated with increased economic consequences for large water storages with a volume higher than 10 Kmc. and with a probability of 8% for seismic events between 5 and 6 magnitude. For medium-size reservoirs (between 1 and 10 Kmc.) the probability of seismic events (M=4-6) has a maximum of only 3 percent.

Smaller reservoirs (between 1 and 0.1 kmc) relate tot he significant probability of smaller earthquakes, between 3% and 6%, for feeble seismic events (between 1 and 2) which are detected by specialized seismic instruments. It should be noted that he same trends are often associated with small reservoirs, where favorable sensitive geological structures may be present (M=4) with 4%, and even higher probability of 6% to 7% are related to smaller seismic events of magnitude two (reservoir smaller than).1 Kmc).

In a prone geological setting, such a probabilistic approach correlated with the concept of hydrofracturation which may develop in any impoundment may be a confirmation to the Rothe's provocative statement "Fill a lake, start an earthquake". The sole amendment is related to the size of such earthquakes which may be more often detectable only by appropriate seismic instrumentation. This can have some significance in dam design and safety concerned zones of low natural seismicity, but may have more importance for seismologists as a potential means of understanding the earth's crust behavior in relation to seismic activities.

The Risk Prediction Model developed in relation to the ICOLD survey, is based on field elements and has the potential of a predictive procedure which could be used to identify associated risks and to develop passive mitigation procedures in relation to the control and design of the rate of impoundment or operation, where significant variations of reservoir level may be possible, e.g. pumped hydro storages.

The best chances to use the model may be in association with actual dam safety concerns in assessing the pertinent input to the Environmental Impact Assessment procedures.

The capabilities of the RPM are still bound to the statistical nature of the data which is used, based on observational and comparative data for risk identification and hazard evaluation. In this line the RPM method should be considered as a specific earthquake prediction procedure related to conception of dam structures.

Considering the much larger domain of earthquake prediction, the RPM should be considered as pertinent to the preliminary statistical stage of seismic prediction techniques.

More elaborate seismic prediction concepts refer to long term predictions based on the strain accumulation stage prior to earthquakes, and more evaluated prediction procedures concerned with generation of the dilatancy stage before seismic events, mainly for medium and short term prediction applications. It could be noted that monitoring of such complex problems has often been related to large programs associated with reservoir monitoring of RIS, e.g. Monticelo in USA.

Recognition should be given to the limited nature of most earthquake prediction programs, which, even with significant seismological advances in understanding, are still far away from the identification of the stages immediately prior to main crust ruptures for practical prediction of short ranges.

Research Requirements for Enhancement of RIS Control

Research requirements for improving control of RIS could not be disassociated from the general requirement of seismological studies for development of earthquake prediction procedures. From this general view point some issues interest in dam safety application may be of greater immediate importance:

1. Evaluation of the long term risk of RIS incidence. This requirement is associated with the long term risks in the development of seismic events on reservoirs such as the Aswan after 17 years, or Marathon, 9 years and case histories disputed with seismic events suspected after 16 years of operations on reservoirs with very small water heads (20 m at Cajun Brazil).

2. Estimation of size of RIS earthquake vis-a-vis of generating more specific parameters required by the dam design procedures, magnitude, accelerations, response spectra, etc.

3. Application of RIS identification and mitigation procedures used for water storages and other industrial activities prone to man-made seismic activities such as: production and re-injection (water flood) of petroleum reservoirs; exploitation of underground steam reservoirs; deep mining operations; emplacement and storage of nuclear explosions. The common element will be significant earth stress modifications.

Conclusions

Evaluation of the specific risks associated with Reservoir Induced Seismicity is possible by using methods such as the Risk Prediction Model including elements of statistical data related to past case histories and demonstrating some predictive capabilities and the potential for development of passive risk mitigation procedures.

The inclusion of statistical components makes the employment of the Risk Prediction Model extremely pertinent to existing Environmental Impact Assessment procedures in which future incidence may be questioned on the grounds of past consideration of the unintentional nature of RIS occurrence for new dam structures. Most effective environmental reduction procedures are related to more in depth estimation of the economic implications of risk components related to new reservoirs, in zones of low natural seismic hazard.

The economic impact analysis of these specific risks are an available tool in reducing or eliminating hazard abatement through Environmental Impact assessment procedures and similar complex analysis, to the financing of such new developments in identified prone areas.

Environmental Impact Assessment (EIA) with full identification of economic implications is the only alternative for the abatement of the environmental hazard of RIS. This option of effective EIA should fulfill the lack of any provision in seismic codes, regulations associated to building or disaster relief procedures related to the risks of Reservoir Induced Seismicity.

PART III

ENVIRONMENT

Chapter 13

THE LARGE DAM CONTROVERSY

W.R. Rangeley

Emphasis on Environmental Factors

Although the need to preserve the quality of the human environment has long since been in the minds of planners of storage reservoirs, the emphasis on environmental factors has become much greater in recent years owing to:

1. Rapid increase in the number of dams, both under construction and planned;

2. The greater population densities of the world, (including in the river valleys where the storage dams are constructed;

3. Scarcity of new land resources;

4. Greater pollution of river waters from agricultural chemicals and sewage;

5. More critical salt balance in river systems;

6. Wider public participation in planning decisions.

Dam designers in the past were required to give careful attention to the aesthetic appearance of both structures and reservoirs. The structure of the dam and its associated works were given architectural features to enhance their appearance and reservoir drawdowns were limited to avoid unsightly bare banks of sediment. Today public tastes have changed and no longer are hydro-electric power stations built to look like cathedrals. Furthermore dams have become larger and it is in any case less feasible to adjust the functional design to provide for architectural features.

Attitudes towards public access to reservoirs have changed and recreational benefits are much more widely exploited.

Of all the issues that arise from reservoir construction the one that stimulates most debate is that of "resettlement" referred to more specifically below.

Classification of Environment Effects

The environmental effects related to reservoir construction may be divided into three broad categories:

1. Effects of the Environment on Reservoirs.

2. Effects of Reservoirs on the Environment.

3. Secondary Effects of Reservoirs on the downstream service areas and downstream systems (i.e. irrigation, navigation, domestic and industrial water supply and power systems).

Figure 1 sets out some of the main effects in each of these three categories.

The Main Concerns of the Environmental Non-Governmental Organizations (NGO)

Throughout the world there are various non-governmental organizations that are devoted to the protection of the environment and to the alleviation of the adverse effects of major infrastructure. Several of them are specifically concerned with large storage dam projects.

The World Bank recently embarked on an exchange of views with a number of NGO's in an endeavour to develop a better understanding of each others motives and points of view. Whereas it is not easy to summarize the concerns of the many NGO's that have different specific areas of interest, the salient points of view presented by a group of NGOs to the World Bank may be described as follows:

1. If there were an accurate accounting of all environmental effects and social impacts, together with a careful evaluation of alternatives, national decision-makers would decide in numerous instances not to build dams. In this context the cost of resettling displaced populations was considered to be the dominant factor.

2. Some NGOs make a general allegation that dam builders are excessively "supply-side" oriented. Much more attention should, in their view, be given to conservation in order to reduce consumption of water and hydro-electric power.

3. The more serious impacts of dam construction that receive inadequate attention are:

    (a) displacement/resettlement of populations

    (b) loss of biological diversity

(c)   health hazards

(d)   waterlogging and salinity

(e)   water quality

(f)   loss of natural areas

(g)   loss of cultural areas

(h)   construction damage

(i)   dam safety

(j)   loss of fisheries

Other than resettlement, the above topics were not discussed in-depth during the exchange of views between the NGOs and the World Bank.

Resettlement--The Big Issue

The main concerns here are:

1. displaced peoples after lack of political influence necessary to protect their interests;

2. compensation for land loss is inadequate. Often the figure is set too low in the first place but even when it reflects market prices initially it becomes inadequate owing to induced scarcity of land resources.

3. monetary compensation, even when adequate, tend to get "lost", either through dissipation on things other than resettlement or through entrepreneurial activities;

4. "land for land" is preferable to "money for land" but not always feasible or legally imposable;

5. There is often insufficient allowance for general damages and "psychological" costs.

Resettlement--Some Precaution

In order to ensure that displacement and resettlement of population is carried out efficiently and fairly it is worthwhile considering the following precautionary courses of action.

1. make the whole process of displacement, land acquisition and resettlement a separate project with funding that is independent of the dam project itself. At the same time care must be taken not to allow abrogation of responsibility by the main project agency;

2. carry out detailed surveys in both displacement and resettlement areas;

3. prepare detailed plans for compensation and resettlement;

4. allow for changes in market prices;

5. discourage entrepreneurial intervention that might deplete the level of compensation reaching the resettlement people;

6. monitor the implementation of the compensation, displacement and resettlement processes;

7. develop long term support program for the resettled;

8. encourage public participation.

These and other recommendations on resettlement project are presented in more detail in World Bank Technical Paper No. 80 "Involuntary Resettlement in Development Projects" by Michael M. Cernea, 1988.

The Case for Conservation

Water Supply

In developed countries savings can be effected by control of garden watering, mandatory economy type toilet flushing system and penal tariffs;

In developing countries it is difficult to consider overall economies in water use when over half of the populations has no systematic supply. Even in many cities half the population are not served by house connections. However most of the above restrictions for developing countries can be applied if needs be. The use of standpipes in urban areas results in big reduction in per capita consumption but is not always socially acceptable.

In most countries urban water system losses can be greatly reduced but because raw water is low, compared with the cost of treated and distributed water, it is often cheaper to supply more water into the system than to renovate leaky reticulation systems. It must however be borne in mind that some 70-80 percent of urban and rural supplies are not lost to the hydrological system of a region unless sewage effluent is discharged to the sea.

Irrigation

Irrigation efficiency in many parts of the world is low; even when treated on a river basin rather than a scheme basis. On a scheme basis it is about 30 percent or less in the vast irrigated areas and Asia whereas 35 or 40 percent should be achievable with some remodelling and improved management. However it is important to bear in mind that most inefficiencies occur in the high flow season and savings made by improved efficiency at such times would need to be stored for use in the low flow season -- calling for more and not less storage dams. In the low flow season, most irrigation systems are water short so farmers do not usually waste water at that time.

Although water consumption is vital in the irrigation sector, the purpose is not to build less dams but rather to increase cropped areas, through both intensification and expansion.

Hydroelectric Power

Many environmentalists allege that energy is wasted and that a "supply side" attitude prevails in the power generating industry. Within the context of dam construction it is important to take the following considerations into account.

The main opportunities for energy savings are in the developed and not developing countries. With few exceptions, such as Canada and the USSR, the potential for future hydro-electric expansion in developed countries is small, if any.

In the developing countries there is still a large unexploited hydro-electric potential and in many cases it represents the only alterative to imported fossil fuel. Even a drastic energy saving scenario such as that presented on "Energy for a Sustainable World" projects a five times increase in hydroelectric generation in developing countries from 1980 to 2020. (In the event it is likely to be much higher)

The Way Ahead

The subject of the environment is generating a wider and deeper public debate relating to all forms of infrastructural development. Storage dams will represent an important element of that debate particularly in developing countries where the number of dams will continue to increase rapidly as demographic growth imposes greater pressures on water supplies. The environmental NGO's and developmental agencies alike must seek for changes of attitude and public strategy if the problems of water supply are to be resolved in an optimal manner. The public strategy must be one of cooperation and not confrontation.

There must be a general recognition by all concerned that those who benefit from dams must generously compensate those who suffer losses as a consequence of their construction.

The survey and monitoring of dams projects for environmental consequences must be carried out in detail at an early stage in the project cycle in order to ensure that there is a full evaluation of environmental consequences--both positive and negative--before a final decision is taken or implementation.

Finally the debate must be based on rational dialectic without the intervention of sophistry.

Figure 1.1

STORAGE DAMS AND THE ENVIRONMENT
CHAIN OF MAIN EVENTS

EFFECT OF ENVIRONMENT ON THE RESERVOIR

EFFECTS	CONSEQUENCES
- Precipitation	- Run-off (for storage) with extreme events of floods & low flows
- Soil erosion	- Siltation of reservoir and outlet blockage.
- Pollutants & natural chemicals	- Deterioration of water quality
- Aquatic life & waterfowl	- Settlement on reservoir
- Evaporation	- Water loss
- Climate	- Low temperature water inflows
- Debris	- Outlet blockage.

RESERVOIR

EFFECT OF THE RESERVOIR AND ITS STORAGE FUNCTION ON THE ENVIRONMENT

EFFECTS	CONSEQUENCES
- Smaller variations in downstream streamflows	- More plentiful and reliable supplies of water; less flood damage; lower after-flood crop production; less flood plain fisheries; more/less estuarial salinity depending on topography. less bank erosion.
- Lower silt content of water	- Lower cost of water management; downstream erosion.
- Inundation of land	- Displacement of settlers; damage to fauna & flora, archaeology, and infrastructure; loss of land.
- Creation of lake/pond	- Recreation; new fisheries; better animal watering but disruption of migration routes; eutrophication.
- Creation of gravity head	- Command of irrigable land; potential power production.
- Lower flows downstream	- More estuarial salinity (in extreme cases).
- Temperature	- Better conditions for users (warmer water, less frazzle ice).
- Interception of river	- Interference with fish migration.
- Inundation of forests	- Poor water quality for potable purposes.
- Induced seismicity	- Induced landslides.
- Reservoir drawdown	- Dry season cropping/grazing.
- Subterranean leakage	- Rise in groundwater.
- Construction activity	- Economic development; environmental change.

SECONDARY EFFECTS ON "SERVICE AREAS" AND SYSTEMS DOWNSTREAM

IRRIGATION

EFFECTS	CONSEQUENCES
- Lower silt content in water	- Lower O&M cost; better water management leading to less water logging.
	- Changes in fauna and flora.
- More regular supply of water	- Adoption of perennial in place of non-perennial irrigation with better crop production.
	- Expansion of irrigated area.
	- Better control of some pests and diseases; less control over others.
- Additional water supply	- More waterlogging but, with good management, better salinity control.
- Gravity supply to irrigation system (where feasible)	- Lower energy consumption.
- Regulated river flows	- Lower pumping costs.

Note: Almost all storage dams supplement water supplies to existing run-of-river systems.

DOMESTIC & INDUSTRIAL WATER SUPPLY

EFFECTS	CONSEQUENCES
- More reliable supplies **	- Less failures and less rationing hence better public health control.
	- Better repulsion of salt intrusion in estuaries giving more sustainable rural water supplies.
- Lower biological quality *	- Higher treatment cost.
- Higher chemical quality *	- Lower treatment cost.
- Wider spatial availability of water **	- Less concentration of urban areas and industries.
- Poorer taste * **	- Higher treatment cost.
- Lower silt contents **	- Lower treatment cost.

Notes: * compared with groundwater
** compared with run-of-river

POWER

EFFECTS	CONSEQUENCES
- Non-thermal energy production	- Displaces fossil fuel and nuclear power and their associated environmental effect.
- Renewable energy source	- Sustainable production.
- Lower cost of production	- Permits special industries such as smelting
	- Provides low cost domestic amenities.
- Simplicity of operation	- Less failure in regions with few skilled human resources.
- Lumpy investment	- Debt burden.

Note: Treats fossil fuel and nuclear power as the alternative sources to hydroelectric (excludes tidal wave, and other sources.

Chapter 14

DAMS AND THE ENVIRONMENT:
RELEVANT IMPACTS

F. Lyra

Natural environment as we know is in constant change. Man has observed many of those changes that happened along the known history of mankind as well as others which took place in geological history and left their traces in structures that can be observed in the present time.

Whenever man enters a natural environment many changes occur either by direct or consequential actions. In modern times man's activity has shown a higher capacity of interference on surrounding natural features.

Among the many structures that men have been creating in the last few centuries dams rank together with cities, industries, roads, airports, along the ones with highest capacity of changing the natural surroundings. Based on study and analysis of past actions and their consequences we can proceed with the construction of new structures, minimizing the harmful consequences of those who have been built in the past.

Science and technology make available to modern man the exploration of natural resources, putting them to the service of society minimizing the risk of damage and destruction. For this purpose there must be a honest analysis of the consequences of previous action, and an earnest effort to minimize harmful consequences. Water is required by modern man in ever growing quantities.

Dams are needed to store water and make it available to populations distributed along the world. Dams are also needed for generation of electric power required by mankind.

Construction of dams can produce changes in the environment that can range from beneficial to disastrous. The study of the ecological equilibrium of the area in which the dam will be built is essential for the appraisal of the consequences of its construction.

The dam itself does not interfere strongly and directly in the environment. The reservoir, however, which is created by the impounding of the water that flowed in the river and will start accumulating behind it, can have strong influence n the environment.

To analyze the environmental impact of the construction of a dam it is necessary to put together the multidisciplinary knowledge of a large group of technicians and scientists. There is, nowadays, a widespread conscience of the importance of this effort, that must be undertaken prior to the decision to build, and be maintained during the implementation of a project.

Many technical organizations are aware of the need to examine the impact on the environment of the works they undertake. The international Commission on Large Dams for instance has had in operation a technical committee especially devoted to the study of dams and the environment. We will review some of the main consequences on the environment of a project which includes construction of a dam.

Impacts on Water

Alteration of River Flows

1. reduction of floods
2. increase of minimum flows

Modification of Water Quality

1. reduction of turbidity
2. reduction of sediment transport
3. modification of the contents of dissolved gases
4. depletion of oxygen - at early stages
5. oversaturation - (low temperatures)

Possibility of Generation of Induced Seismicity

Eutrophication of Water Bodies

Evaporation versus Evapotranspiration

Reservoir Stability and Riverbed Erosion

Impacts on vegetation

Loss of Individuals within the Reservoir Area

Modification of Plants Habitats

Impacts on Fauna

Ichthyofauna - Consequences of Alterations of their Living Media

1. lentic fish (still waters)
2. lotic fish (running waters)
3. migratory fish
4. semiaquatic mammals

5.   water mammals
6.   reptiles
7.   water fowl

Impacts on Human Activities

Displacement of Populations

Modification of Population Activities

1.   extractivist
2.   agricultural
3.   urban occupations
4.   transportation means and facilities

Reduction of Agriculture and Cattle Raising Lands

1.   increase of irrigation possibilities
2.   potential increase in fish production

Flooding of Archaeological and Cultural Sites and Landmarks

Impacts on Public Health

Waterborne Diseases

Multiplication of Vectors

1.   insects
2.   mollus

Chapter 15

ENVIRONMENT AND IRRIGATION

Montague Yudelman

For the record, I was the Director of the Agricultural and Rural Department at the Bank from 1972 to 1983, a period in which there was a very rapid expansion in Bank lending for irrigation, especially in Asia. Since leaving the Bank, I have been associated with two highly regarded environmental organizations: the World Resources Institute and The Conservation Foundation. My recent experiences have given me an opportunity to learn something about the concerns of all kinds of professionals involved in the debate about the role of irrigation in sustainable development. There is a minority that argues very vigorously that the environmental consequences of the spread of irrigation have been so bad that there should be a shift away from irrigation, especially by large dams, to meet future needs for food and fiber.

They argue too that investments in irrigation involve the inefficient use of capital and that these investments generally benefit the rich at the expense of the poor. However, I believe that a vast majority of professionals --and I include myself among them--believe that irrigation has reduced poverty (though it has worsened income distribution). More significantly, though, most believe that there is no other way to meet the needs of a population growing by one billion people each decade, without irrigated agriculture continuing to play a major role.

At the same time, I also share the widely held view that the expansion of irrigation will have to be much more sensitive to environmental concerns than in the past and that concern for the environment must be seen as an integral part of the process of change and not just as an "add-on" to a development effort.

In this context, I would like to mention two rather general global issues before discussing project-related issues. The first is whether there will be enough fresh water in the world to meet the needs of seven billion people or so by the year 2010, just 20 years away, an addition of more than two billion (mostly in the poor countries). this is especially relevant given the fact that irrigation is estimated to use 90 percent of the available supply of fresh water at a time when there is a global explosion in urbanization, and rapidly growing non-agricultural requirements. Studies, including those commissioned by the Bank, indicate that there should be enough fresh water to meet global needs, but water supplies in many areas are being drawn down as localized competition increases for available supplies. How should this competition be resolved? One suggestion is that the market should determine allocations.

This approach is widely favored by production economists especially as it seems that higher priced water would reduce wastage and increase productivity. This may well be so, but most environmentalists seem to lean very heavily in favor of regulation of such things as uncontrolled pumping rather than relying on the market as a vehicle for safeguarding the sustainable supply of this renewable resource. I think both are needed. Incidentally, rational use of global supplies would undoubtedly benefit from multinational riparian agreements--a special problem in Africa.

A second global concern is the relationship between irrigation, the greenhouse effect and climate change. This is a vast topic in itself; environmentalists seem to be less skeptical about the effects of man-made pollutants on the atmosphere and climate change, than do others. Consequently, they favor energy policies that reduce the emission of carbon dioxide; such policies could well include the expansion of large-scale hydroelectric power and multipurpose water projects. I don't know whether this should be taken into account when considering the economic and social viability of an investment in a large multipurpose project. Perhaps it should; however, this creates a dilemma for many environmentalists who are against large-scale projects because of their unanticipated--and in their view usually harmful--effects on the environment. Maybe there should be a reconsideration of the potential of small-scale hydropower.

There are a number of environmental concerns that relate to many irrigation projects per se that are also of concern to production specialists. By far the most significant of these problems relate to salinization and waterlogging with the near-destruction of once productive land. The acreage involved are staggering; one of the higher estimates is that forty percent of all irrigated land has been adversely affected. The costs of reversing this degradation are high and the losses in production have been substantial. What baffles most environmentalists concerned with improved natural resource management is why salinization has continued, almost unchecked, for so long when specialists have long known how to limit its spread. They argue that the long-term conservation of resources has been given very low priority compared with that given short-term increases in output. Perhaps they are right, but it seems to me that there is no reason for these earlier priorities to continue.

Two other project-related issues are familiar to you; they are involuntary resettlement of people when land is to be inundated, and health concerns. Need I stress that the displacement of poor people is always a matter of concern, but it becomes even more troublesome when it involves moving these people to ecologically sensitive areas such as mountain slopes of forest reserves. Similarly, the spread of waterborne diseases following on the introduction of irrigation is a matter of concern. Experience, such as with the development of the Aswan Dam, seems to indicate that this is best contained by public health programs operated in tandem with irrigation programs. This still seems to be a relatively rare phenomenon, and I look forward to hearing what Mr. Olivares has to say on this subject.

There are other environmental issues such as the rapid siltation of reservoirs, usually arising from poor watershed management; then too there are downstream problems arising from the intensification of agriculture <u>following</u> on the introduction of irrigation. These problems range from the fears of vulnerability from the genetic uniformity of high yielding varieties to the problems associated with overuse of agrochemicals, especially pesticides. However, these problems are common to nearly all efforts to intensify agricultural production and, as such, extend beyond concerns with irrigation *per se*. The resolution of these problems is part of the more general problem of managing the whole agricultural sector and promoting environmentally sound agricultural development by such steps as removing subsidies on pesticides, banning the use of certain pesticides, developing new varieties of see and the like.

There are a number of steps that can be taken to promote environmentally sound irrigation projects. These include paying more attention to upstream and downstream facets of projects, e.g., rivershed control to limit erosion and control of downstream riverbank erosion. In the main, all these approaches involve a more comprehensive view of what constitutes irrigation projects than hitherto, with greater contributions by a wide range of specialists, many of whom are not usually consulted in irrigation project development, e.g., health specialists, ecologist, foresters, anthropologists, marine biologists and so forth.

Another needed step is to have an improved methodology for estimating and measuring the benefits and costs from some of these added interventions. At the very least the methodology should be able to indicate the order of magnitude of environmental costs and benefits from adding or dropping components so that policy makers can use this information to help them make their decisions.

Many environmentalists believe too that another needed step is that analysts should give <u>explicit</u> weight to smaller rather than larger projects where such options do exist. Smaller projects are usually easier to manage, an important consideration in may countries, and they involve less in the way of sunken costs than do larger projects, so that there is less of risk if things go wrong. In addition to these consideration, steps have to be taken to give due regard to the more familiar issues that pertain to sustained development of irrigation: more investment in maintenance, involvement of local groups of water users, the levying and collection of water charges and the provision of adequate funds for upgrading the management of systems.

In conclusion, let me say that I believe that irrigation will necessarily continue to be a major engine of growth for agriculture for the foreseeable future. However, I also believe that experience shows that the design and implementation of environmentally sound irrigation projects will require more and better basic data and that preparation, planning execution and implementation will take longer than did the projects of an earlier period. These projects, being more comprehensive, will also be more costly per hectare at a time when capital is becoming increasingly scarce, and costs of irrigation per hectare are rising.

Thus an important challenge for the future lies in looking at environment and production options and striking the best possible balance

between them. There will <u>have</u> to be a compromise, especially over sequencing
of irrigation, because the resources simply will not be there to satisfy all
needs; finding the right compromise will involve many of the issues which, I
am sure, will be raised in today's discussion.

Chapter 16

RESERVOIR WATER QUALITY AND MANAGEMENT

Ralph H. Brooks

Impoundment of a flowing stream to form a reservoir causes many changes in the physical, chemical, and biological characteristics of the water resource. The specific nature and extent of these changes vary, depending on climate, geology, topography, length, depth, volume, retention time, orientation to prevailing winds, straight or serpentine configuration design and location of operations, quantity and quality of inflows, and other factors. However, the general nature and extent of these changes are invariably such that the water resource can no longer be considered a stream. Neither can it be considered a lake. Although a reservoir is in some ways similar to a stream and in other ways similar to a natural lake, it is also different in many ways from either of these and must be considered separately as a third basic type of water resource system, with its own set of inherent characteristics and behavioral phenomena.

The effects of a single impoundment on the resulting reservoir and tailwater stream are complex. Effects of multiple impoundments are more complex. Characteristics of releases from one reservoir can determine or strongly influence characteristics of the next reservoir downstream. Characteristics of reservoirs cannot necessarily be evaluated by using the same standards and criteria developed for application to streams or natural lakes. Their desirability must be evaluated by weighing and balancing the effects of changes on the intended uses of the new type of water resource that has been created.

All desired uses of the resource may not be consistent with the inherent capabilities of the modified system created by impoundment, and requirements for some desired uses may conflict to varying extents with requirements for other desired uses. Therefore, value judgments of effects of impoundments must be based on appropriate expectations and criteria related to designated uses compatible with the capabilities of this type of water resource. When more than one use is designated for a reach of a reservoir, compromise may be necessary.

The TVA Integrated Multipurpose Reservoir System

Components

The Tennessee Valley Authority (TVA) was charged by the TVA Act of 1933 to develop and manage the resources of the 106,00-km2 drainage basin of the Nation's fifth largest river, a region encompassing parts of seven southeastern States. TVA development of water resources has included building or acquiring 36 dams that now comprise an integrated regional water control system (shown in figure 1 and characterized in table 1). There are 9 multipurpose dams on the mainstem Tennessee River; 26 dams on Tennessee River tributaries, including 13 multipurpose dams (storage dams), 7 single-purpose dams (power dams that as far as practical are also now operated to provide other benefits to the system), and 6 "tributary area development" dams (that were built primarily to provide local development benefits such as water supply and recreation but that also provide some benefits such as power generation or flood control); and 1 single-purpose (power) dam on a Cumberland River tributary.

TVA also has built a pumped-storage project and 10 small tributary area development dams that are not considered part of the integrated water control system. Another tributary area development dam is partially completed. There are also an additional 28 publicly or privately owned dams within or bordering the region that are not now operated as part of the TVA integrated water control system. An open navigation canal connecting Kentucky Reservoir on the Tennessee River system with the Corps of Engineers Barkley Reservoir on the neighboring Cumberland River system allows free exchange of water and biota between these two systems. The recently completed Tennessee-Tombigbee Waterway connects Pickwick Reservoir on the Tennessee River system with the Tobigbee River system and ultimately the Gulf of Mexico by means of a canal and locks; it results in a net export of water from the Tennessee River system.

Primary and Secondary Objectives

TVA operates the integrated water control system to provide multiple regional benefits. The TVA Act established operating priorities for navigation, for flood control, and as far as consistent with these purposes, for generation of hydroelectric power. Within the limits of these constraints, the system is also managed for enhancement of water quality, stream and reservoir recreation, fish and wildlife resources, and water supplies for farms and cities and industries; assimilation of treated waste effluents; emergency dilution of accidental pollution spills; augmentation of low flow; control of aquatic vegetation and mosquitoes; accommodation of construction and other activities along shorelands; and other special purposes in the regional public interest.

Conflicts Among Purposes

From a regional and long-term perspective, operation of the integrated multipurpose water control system provides significant benefits for all of these aspects of resource development, use, and conservation. In local areas at certain times, however, system operations can adversely affect or preclude one or more desirable uses. For example, flood-control operations during a wet spring can preclude stabilizing reservoir water levels to aid fish spawning.

Water released downstream to dilute pollution from accidental spills is no longer available to serve upstream uses. Systemwide operations for the regional priority purposes of commercial navigation and flood control benefit primarily the downstream areas along the mainstem reservoirs; however, releases from tributary reservoirs to maintain downstream flows during the dry season or to provide storage capacity during the flood season require wide fluctuations in tributary reservoir water levels and tailwater flows. These fluctuations limit various types of lake and stream recreation, which is a secondary priority from a regional perspective but often the highest priority from a local perspective. Lake users generally desire lake levels to remain high and stable as long as possible each year for their recreational purposes. Some types of tailwater recreation such as whitewater rafting and canoeing require lake-lowering releases to provide high flows. Other types of stream boating and bank or wade fishing require low flows.

In response to evolving regional needs, TVA is now conducting a formal reassessment of its reservoir system operating policy. After numerous public meetings and other surveys to determine user needs and desires, TVA is attempting to determine if the priorities established more than a half century ago still server modern needs or if a new set of priorities should be adopted by amending the TVA Act, if necessary. If change is determined both desirable and feasible, finding the proper balance among the often conflicting uses and functions of the reservoir system and at the same time ensuring protection of environmental quality will involve many difficult decisions and will require cooperation and compromise among all users.

System Characteristics

Many of the conflicts among multiple uses of the sam e resource are attributable to inherent differences among the subregions within the Tennessee Valley and among the types of projects that comprise the integrated system.

Mean annual rainfall across the region (1980 to 1988) is 130.3 cm, but rainfall is not always average, and it is not distributed equally across the region or throughout the year. Annual rainfall has varied from a minimum of 90.9 cm to a maximum of 165.4 cm. Prolonged drought during recent years has resulted in a cumulative regional deficit of about 152 cm (June 1984 to February 1989), or a deficit of more than a year's worth of average annual rainfall. The western half of the Valley downstream from Chattanooga, Tennessee, receives slightly more rainfall than eastern half upstream from Chattanooga (130.2 versus 128.6 cm); however, some areas in the mountainous eastern third of the Valley receive an average of about 228 cm annually. More than half the total rainfall is received from December through mid-April, while September and October are usually very dry months. Mean annual runoff for the Valley (1975 to 1988) is 56.5 cm, but extremes in dry and wet years have varied from 26.6 to 87.3 cm.

These differences in rainfall and runoff, combined with differences in topography and other physical factors and with different sets of multiple objectives for water resource management in different parts of the system, require different types of reservoirs and reservoir operations in different parts of the region.

Tributary Projects

Most tributary multipurpose storage impoundments are located in roughly the eastern third of the Valley. This subregion is generally dominated by rugged, forest-covered mountains, which in some areas receive high rainfall and yield high runoff. Nolichucky is filled with silt from mine erosion, and Melton Hill and Tellico are connected to the mainstem navigation system and have some features and operating characteristics of both tributary and mainstem reservoirs. The other 10 reservoirs in this group have maximum depths at the dams ranging from 38 to 141 m and averaging about 66m; lengths ranging from 21 to 208 km and averaging 62 km, and volumes ranging from 210 to 2,516 Mm^3 and averaging 1,060 Mm^3.

As might be expected, hydraulic retention times at these 10 storage projects tend to be very long, ranging from 37 to 400 days and averaging about 216 days (based on normal maximum pool volumes and average annual streamflows). Seasonal differences between normal minimum and maximum water levels tend to be extreme, ranging from about 13 to 39 m and averaging 18 m. Turbine intakes tend to be located deep in these reservoirs to permit hydroelectric generation throughout the annual cycle of high and low water levels. Five of these projects have bottom-level intakes, and five have midlevel intakes (although even midlevel is deep in these deep reservoirs).

The tributary single-purpose projects were acquired or built primarily to produce electric power. Their characteristics vary. Like the tributary multipurpose projects, seven of the eight tributary single-purpose projects are located in the eastern highlands on headwater streams of the Tennessee River system; one, Great Falls, impounds a tributary of the Cumberland River. Apalachia, Fort Patrick Henry, and Wilbur are regulating flow-through projects downstream from large multipurpose tributary projects, intended primarily to reuse their discharges for generation of additional power. Only Ocoee No. 1, Blue Ridge, and Great Falls have significant storage capacity. Ocoee No. 2 is essentially only a diversion dam, directing flow through a flume to a downstream powerhouse.

Ocoee No. 3 is almost filled by silt from erosion of a large area denuded a century ago by mining, clearcutting, and smelter gases. Reservoir lengths range from 3 to 35 km and average about 15 km. Depths range from 7 to 37 m and average about 27 m. Retention times range from a few hours at Ocoee No. 2 and Wilbur to 157 days at Blue Ridge. Fluctuations between normal maximum and minimum water levels range from 1.5 to 30.5 m.

The six tributary area development dams were built primarily for local development purposes such as water supplies and recreation, but they also provide system benefits. All provide flood-storage capacity, and one, Tims Ford, generates power. Tims Ford on the Elk River and Normandy on the Duck River are in the rolling farmlands of middle Tennessee. The other four impound Bear, Little Bear, Upper Bear, and Cedar Creeks in the extreme northwestern corner of Alabama and are collectively known as the Bear Creek projects. Reservoir lengths range from 10 to 55 km, depths range from 9 to 39 m, retention times range from 3 to 28 days, and fluctuations range from 3 to 7 m.

Mainstem Projects

The nine mainstem multipurpose impoundments lie west of the mountains in a broad, crescent-shaped basin sweeping southwest out of eastern Tennessee into northern Alabama, then west into Mississippi, and finally north through western Tennessee and into Kentucky to the Tennessee River mouth on the Ohio River. Rolling hills with scattered farms and upland woods dominate the upper reaches; expansive flatlands of farm fields and hardwood bottoms characterize the lower reaches. Mainstem reservoirs tend to be shallower and longer than the tributary storage reservoirs, although volumes are generally similar.

Flows are much higher, and retention times are much shorter depths at the dams (deep water in the original river channel, flanked by wide, shallow, "overbank" areas) range from 20 to 34 m and average 25 m; lengths range from 25 to 297 km and average 119 km; volumes range from 299 to 3,346 Mm3 and average 1,174 Mm^3. Retention times tend to be very short, ranging from about 5 to 34 days and averaging 16 days (based on normal maximum pool volumes and average summer stream flows). Seasonal differences between normal maximum and minimum water levels are limited by navigation requirements, ranging from 0.6 to 2.3 m and averaging less than 1.5 m. Five dams have bottom-level turbine intakes, and four have midlevel intakes.

Annual Operating Cycles

Annual operations of the tributary multipurpose (storage) reservoirs are keyed to the annual cycle of rainfall and runoff and to the system functions these projects serve. Generally, as the probability of flood-producing levels of rainfall decreases and as regrowth of runoff-retarding vegetation increases in the spring, tributary reservoirs are filled gradually and, if rainfall permits, reach their normal maximum levels by early summer. The stored water is gradually released through the dry late summer and fall. By early winter (the first of January) tributary reservoirs reach normal minimum levels that provide storage capacity for possible floods during the late winter and early spring.

Mainstem reservoirs also operate on an annual cycle, which incorporates additional features in response to their inherent characteristics and the multiple primary and secondary purposes they serve. In early April, they are filled as rapidly as available flow permit to normal maximum level, or sometimes slightly higher for a brief time to strand floating debris that might provide habitat for mosquito production or pose other problems. They are maintained at full-pool level until late June. Eight reservoirs then begin weekly 0.3-m fluctuations to disrupt mosquito production. On four of these fluctuation is combined with a gradual recession of about 0.03 m/week.

Kentucky Reservoir, too large to permit rapid fluctuation, is operated more like the tributary storage impoundments, with only a gradual recession during this period. Drawdown is accelerated in the fall, and normal minimum pool is reached by the end of the year (of course, drawdown releases from all tributary reservoirs must also pass through the mainstem reservoirs during this period). Except during flood-control operations, minimum pool is maintained until spring, when, with the end of the flood season, the cycle begins anew. The extent of drawdown on these reservoirs is limited by the requirement that they provide a minimum navigation depth of 3.4 m.

Effects of Impoundments on Flows, Water Quality, and Ecology in the Tennessee Valley

Beneficial Effects

Much attention is focused on assessing and managing various adverse effects of impoundments. However, any discussion of the effects of impoundments would be incomplete and misleading if it did not emphasize that most effects are beneficial.

Flow

The TVA reservoir system transformed the Tennessee River and its tributaries from barriers to development into one of the most controlled and useful river systems in the world. Average annual discharge at the mouth of the Tennessee is about 1,800 m3/s. Before regulation, extreme flows there varied unpredictably from a trickle of 127 m3/s during the drought of 1925 to a torrent of 13,400 m3/s during the flood of 1897. The general effect of operations of the completed reservoir system has been to moderate extreme flows by augmenting the level of low flows and reducing the level of peak flows.

Although system operations result in numerous brief periods of no flow below many dams, they increase the general level of low flow throughout the full length of the Tennessee River. Except during periods of no flow to reduce flooding on the lower Ohio and Mississippi, minimum flow at Kentucky Dam near the river's mouth is now usually regulated to achieve appreciably higher flows than those would occur naturally.

Natural flow during the driest three months of a typical year averages about 620 m3/s, but regulation almost doubles that flow to 1,050 m3/s. Comparison of 7-day, 10-year minimum flows (flow, averaged over a continuous period of 7 days, that occurs with a frequency of one year in 10 years) at Whitesburg, Alabama, before and after construction of most of the TVA dams also indicates doubling of low flows about midway between the mouth and head. Minimum flow there from 1925 to 1935 (before completion of any TVA dam) was about 187 m3/s; that for 1945 to 1963 (during which most of the large upstream TVA dams were in operation) was about 410 m3/s. Mean flow at Whitesburg for 62 years of record (1924 to 1986) is 1,217 m3/s. Because of regulation provided by the upstream tributary reservoirs, these relationships also hold true at the head of the Tennessee River.

Before completion of the reservoir system, low flows and treacherous shoals barred navigation. Limiting depth from the mouth of the Tennessee River at Paducah, Kentucky, to the existing Wilson Dam was 1.2 m; that from Wilson Dam to the head of the river at Knoxville, Tennessee, was 0.5 m. The nine mainstem dams and locks now provide a year-round minimum navigation depth of 3.4 m for this entire 1,040-km reach, connecting 177 public and private barge terminals in this region with the Interconnected Inland Waterway System that leads to ports in 21 States and ultimately to the Gulf of Mexico and ocean ports throughout the world.

At the other extreme of flows, the system does not seek to eliminate all flooding but to manage peak flows to provide a high degree of protection during floods. From 1936 through 1984, system operations had prevented cumulative flood damages of more than $2.6 billion at Chattanooga (the most flood-susceptible urban area in the valley), about $262 million at other areas in the valley, and about $137 million along the lower Ohio and Mississippi Rivers.

Taking advantage of river control in another way, the 29 TVA dams with hydroelectric generating facilities now have an installed capacity of almost 3,330 MW and in a normal year generate 14,114 GWh. In addition to serving well these primary purposes, the TVA system of impoundments also benefits many secondary purposes involving water supplies, water quality, and aquatic life.

Water Supply

The Valley experienced severe drought during the last four years, and several water supply systems that depend on groundwater or unregulated small streams suffered severe shortages. However, none of the water supply systems that draw from the regulated surface waters of the Tennessee River or its major tributaries experienced any shortage because of source limitations. Minimum flows are provided to benefit municipal and industrial water supplies below Tims Ford, Watauga, South Holston, and Normandy Dams. An industrial plant compensates TVA for special water supply releases from Fort Patrick Henry Dam.

Chapter 17

HEALTH IMPACTS OF IRRIGATION PROJECTS

Jose Olivares

Introduction

"Too often, in water resource management schemes designed to advance
agricultural production in arid lands, local populations as well as resettled
people are neglected in the process, and they derive little or no benefit to
their health. As with other aspects, the impact of irrigation on human health
has some serious deleterious effects; it also has important positive
effects".[14] "The construction of dams, formation of man-made lakes and
development of irrigation projects in tropical areas introduce important
changes in the environment and, in parallel, produce a number of risks to
human health, apart from the evident benefits brought to the macroeconomy of a
country or the production of some specific economic advance."[15]

Such statements capture some of the most important public health aspects
of irrigation projects. Irrigation is the largest single component of
multilateral lending for agricultural development. It is an ancient art which
provided the agricultural basis upon which many early civilizations, such as
in China, Egypt, India, Mesopotamia and Peru developed and flourished.
Despite its ancient history, irrigated agriculture is increasing at a rapid
rate throughout the world, registering a 21.7 percent increase between 1961
and 1976.[16] And it still has much scope for growth.

Public health impacts of irrigation projects can be both positive and
negative. Unfortunately, many recent irrigation projects, especially in
Africa, have spread or amplified many water-related human diseases. The
article by Hunter et.al. (op. cit) should be read in full for a comprehensive

14. Worthington, E. Barton (editor) 1977. Arid Land Irrigation in Developing Countries:
    Environmental Problems and Effects. Based on the International Symposium, 16-19th February,
    1976, Alexandria, Egypt. Pergamon Press, Oxford.

15. Hunter, J.M., L. Rey, and D. Scott. "Man-made Lakes and Man-made Diseases:
    Towards a Policy Resolution", Social Science and Medicine, Vol. 16, p. 1127-1145, 1982.

16. Berry, L. The Impact of Irrigation on Development Issues for a Comprehensive Evaluation
    Study. Discussion paper prepared for U.S. Agency for International Development, Program
    for International Development, Clark University, Worcester, Massachusetts. 1980.

look at this issue. Another article, by Rosenfield and Bower,[17] discusses
this area thoroughly also, although it is restricted to only one disease,
schistosomiasis, in expository detail. The detail is, however, _rewardingly
great_.

Major Characteristics of the Methodology

The methodology provides a method for assessing _ex ante_ the public health
risks associated to any irrigation project. It attempts to provide the World
Bank and its member governments with a framework that allows irrigation
projects to be appraised for their public health impacts as they are being
considered on their technical and economic merits. Five salient features of
the methodology are described below.

Specific Project Linkage. Each river basin, each project area is
independently assessed for its unique characteristics. These define the
public health risk that will be made explicit as the output of this exercise.
This, in turn, provides decision-makers with adequate information to better
guide their project selection from the available alternatives.

Shortness. The time and work required to utilize the methodology
are intended to be short: some two expert-months per country. Within these
two expert-months, one might assume a preparation time of approximately two
weeks (to allow for preliminary background research); three weeks of in-
country time (during which project-specific information is collected); another
two weeks for report writing, and an additional week to be used as flexible
reserve.

Cost-effectiveness. This term is used here not in rigid economic
sense but in the more general sense of "optimal ignorance" and "proportionate
accuracy." Optimal ignorance is taken to mean deciding in advance what _does
not_ need to be known, thereby freeing oneself from the (time-consuming) task
of acquiring unnecessary information. Proportionate accuracy is a facet of
optimal ignorance, i.e., if knowledge of the order of magnitude or direction
of change is all that is necessary or useful, then precision to two decimal
places, or at the 95 percent confidence level, would constitute a waste of
valuable resources.

Emphasis on Utilization of Existing Information. Most of the
information necessary to make an assessment of the health risks involved in a
specific irrigation project already exists and is readily retrievable.
Primary data collection is, for the most part, unnecessary, but some overt
gaps in the existing information may be encountered. Minor information gaps
can be tolerated. When such gaps are found, they must be made explicit and
clearly labelled as such. Extensive information gaps, however, are another

17. Rosenfield, P.L. and B.T. Bower, "Management Strategies for Mitigating Adverse Health
Impacts of Water Resource Development Projects, in _Progress in Water Technology_,
Vol. 11, Nos. 1/2, pp. 285-301, 1979.

thing. If they are sufficiently extensive, the validity of the entire health impact assessment exercise should be called into question. Identification of research priorities (as opposed to investment options) in irrigation would then result.

Emphasis on Minimizing Biases. Information should be acquired with conscious attention toward minimizing the biases of what has been referred to as "rural development tourism".[18] These biases favor the accessible areas over the inaccessible, and the (relatively) prosperous areas over the poor (as applied to both environment and people). In spatial terms, the urban over the periurban, the periurban over the rural, the tarmac and the roadside over the areas far from roads, and the center as opposed to the periphery of villages. In terms of persons, contacts with the rural elite may inhibit contact with actual target population. Men relate better to other men than to women. Consultants and project officers tend to see adopters more than non-adopters, and tend to relate more to users of services than to non-users. In terms of seasons, missions tend to cluster during the healthier, food-abundant dry seasons and stay away from the rural areas in the hunger-and-illness-prevalent wet seasons, when travel is difficult.

Diseases to be Considered

Some 30 water-related diseases have been linked in some way to irrigation projects, and could be considered in some detail within the context of this methodology (See Table 1). (Water-washed diseases, unlike the other three categories, may be expected to _decrease_ if the project improves water accessibility, quality or hygiene). While some distinctions can be made in terms of importance or relative priority, there are no universally accepted criteria to choose among them the "most important" ones. Particularly, some emphasize the endemic, widely-spread, debilitating chronic diseases, while others give a higher priority to the epidemic, life-threatening acute diseases. A summary view of both alternatives is provided below; some criteria to select the diseases to be analyzed in a particular case are then discussed.

On the endemic, chronic side, four diseases are the most important ones: (i) malaria, (ii) onchocerciasis, (iii) schistosomiasis (or bilharziasis), and (iv) filariasis. They derive this special status from: (a) the high degree of disability and/or fatality they engender in the individual with the disease; (b) the large proportion of any population exposed to their risk that actually becomes ill; (c) the difficulty and expense of dealing effectively with these diseases once established in a human population; and (d) the fact that they are not occasionally epidemic or with single episodes that yield substantial

---

18. See The Workshop on Rapid Rural Appraisal, October 1979 and the Conference on Rapid Rural Appraisal, December 1979, both at the Institute of Development and Studies, University of Sussex, Brighton, in _Agricultural Administration_, Vol. 8, No. 6, 1981.

immunity, but are, instead, grim burdens that weigh heavily on many aspects of daily human existence.

Table 1

Selected Water-related Diseases And Pathogenic Agent

Category	Disease	Pathogen (vector)
Water-borne	Amoebic Dysentery	protozoa
	Ascariasis	helminth
	Bacillary dysentery	bacteria
	Cholera	bacteria
	Diarrhoeal disease	miscellaneous
	Enteroviruses (some)	virus
	Gastroenteritis	miscellaneous
	Infectious hepatitis	virus
	Leptospirosis	spirochaete
	Paratyphoid	bacteria
	Typhoid	bacteria
Water-washed	Conjunctivitis	miscellaneous
	Leprosy	bacteria
	Relapsing fever	spirochaete
	Scabies	miscellaneous
	Trachoma	virus
	Typhus	rickettsiae
	Yaws	spirochaete
Water-based	Chlonorchiasis	helminth (snail, fish)
	Diphyllobothriasis	helminth (copepod, fish)
	Fasciolopsiasis	helminth (snail, plant)
	Dracunculiasis	helminth
	Paragonimiasis	helminth (snail, crab)
	Schistosomiasis	helminth (snail)
Water-related Insect vectors	Arbovirus (some)	virus (mosquito)
	Dengue	virus (mosquito)
	Filariasis	helminth (mosquito)
	Malaria	protozoa (mosquito)
	Onchocerciasis	helminth (Blackfly)
	Trypanosomiasis	protozoa (Tsetse fly)
	Yellow fever	virus (mosquito)

Source: Adapted from Feachem, R., M. McGarry, D. Mara (eds). Water, Wastes and Health in Hot Climates, John Wiley & Sons, London, 1977.

Among the epidemic, acute diseases, the most important ones are cholera, typhoid fever, and the diarrheas. Their episodic appearance may terrify whole populations (e.g., cholera); the mortality rate associated with such episodes may be high, and they are explicitly perceived as a threatening disease by the affected populations and public authorities.

No more than three or four diseases should be analyzed in a given country study. The selection among them should be made in accord with the national authorities. Several elements should be considered in marking the selection; two among them deserve an explicit discussion: (i) the relative importance of each disease, and (ii) their association with irrigation development.

The main quantitative indicators which can be used to compare the relative importance of different diseases are prevalence, incidence, and the number of healthy days of life lost. They are briefly discussed below (paras. 23 to 27). They do not provide a universal ranking, however, because some are better to measure some kinds of diseases while others better reflect the situation of others. One is required to interpret the available data. For example, cholera is usually experienced by only a small percentage of a cholera-prone populations, lasts just a few days in its victims, is not usually known for second attacks in the same person, and generates more survivors than fatalities; impact may be of a smaller order of magnitude than that of the chronic diseases mentioned above, each of which may easily consume many years of an individual lifetime.

Regarding their association with irrigation development, the methodology deals only with incremental risks, that is, with the increase in prevalence, incidence, or number of lost days which are strictly attributable to the addition, of an irrigation project to a given landscape. The background disease level, as well as the irreducible disease problems/accidents inherent in any kind of development project, are not the subject of this methodology. But another decision has to be made here: irrigation development adds both water and people to certain areas. Under the Study, it was decided to restrict the analysis to those diseases which are water-related (see para. 10). But it is also possible to consider other diseases, which are rather related to the immigration/migration patterns, to the increased population density, and to the increased "urbanization" caused by the development resulting from irrigation rather than from irrigation itself.

Answers to be Provided

The actual aim of the evaluation is to assess what health risks can be expected _if_ an irrigation project is added to a given landscape. This requires sifting through the mass of health and health-related data to identify those few factors which would really make the difference for each particular site where irrigation projects have been identified. In particular, the analysts should ask themselves questions such as: (a) would the introduction of irrigation create favorable conditions for the multiplication of a pathogenic agent, or of a vector, which is already present; (b) would it create conditions for the introduction in the areas susceptible of irrigation of new pathogens or vectors, for the disappearance of either, or for their replacement by others; (c) would it induce population movements which might bring into the area to be irrigated new pathogens or vectors; (d) would it induce behavioral changes (e.g.: regarding domestic water supply, laundering, bathing, swimming, child playing, waste disposal, etc.) which would modify the relationships between people and the environment (water contacts, vector contacts, farming practices, housing). Information on water availability in the different seasons; location of human dwellings vis-a-vis irrigation canals and irrigated areas, and the like should be of help.

One piece of information which is particularly useful is the main characteristics of the vectors and their environmental requirements. Ideally, one could draw the line between the areas at risk and those which would be safe if the limits for vector reproduction, feeding, etc. could be established. There are some examples in this regards: (a) in the Tadla area, in Morocco, there is schistosomiasis in the right bank, irrigated with river waters, and there is none in the left bank, irrigated with cooler mountain waters; water temperature has been identified as the critical factor which makes snail population possible in one area and impossible in the other; (b) in the Philippines, there is schistosomiasis in the westward watersheds and not in the eastward ones; water pH has been identified as the crucial factor; and (c) in Puerto Rico, a snail transmitting schistosomiasis was found to be associated with a particular plant (Caladium supp.), whose habitat could be identified by satellite.

Although some similar work has been done also for mosquitos, black flies, tsetse flies, and other vectors, the body of information available in this regards is not large enough for the Study's purposes. Firstly, most information available refer to the typical, or normal, characteristics of the environments where vectors thrive, rather than to the extreme values which define the border conditions for their survival. Second, different vector species, with different requirements, may fill the same niche in the pathogen's life/transmission cycles; the limits, or critical values, should be found for them all. Besides, vectors, as any other biological material, display large individual variations and a great capacity to change to adopt to environments different from those which are known to be their "normal" ones. Thirdly, what determines the presence or absence of a vector in a certain landscape is not usually a single factor but rather a combination of them, most of which are not themselves critical but that together add up to a favorable or otherwise situation.

Ideally, the analyst should aim at defining the likely health risk at five levels:

1.   areas already infected with high levels of transmission;

2.   areas already infected with modest prevalence where irrigation could lead to epidemic or hyper-endemic situations;

3.   areas currently without disease that could become highly receptive to disease after irrigation;

4.   areas currently without disease that could become marginally receptive to disease after irrigation development; and

5.   risk-free areas.

In practice, it proved difficult to discriminate between levels 2 and 3 in most cases; in Morocco only two levels (at risk; risk-free) could be assessed. Again, maps comparing the project areas with adjacent areas will be useful in establishing a regional in-country perspective. If important zoonoses, such as bovine schistosomiasis, need to be considered, similar maps should be prepared for these diseases.

Outputs

It is now useful to describe the outputs expected from a country study, in order to link the data collection phase to specific outputs, and to avoid wasting time in getting information which is not required or is in the wrong form. The major anticipated outputs can be grouped in three sub-sets: (i) maps, (ii) quantifications of health impacts, and (iii) actions recommended.

Maps

Two sets of maps should be prepared for each of the diseases being analyzed. The first set would describe the current distribution of each disease, a previous distribution--say 5 or 10 or 15 years ago (whatever conveys the most meaningful trend information)--as well as the range or distribution of the disease in adjacent countries. There must be separate maps for vector (if there are any) and for pathogen agent. It will usually be noted that the vector (the snail for schistosomiasis, the mosquito for malaria, etc.) is usually distributed much more widely than the disease. This will define an area of potential transmission. Such transmission would then occur if and when the parasite were introduced to the project area by someone working on the project, or in some other way. The second set of maps would define the risk zones as described in para. 19 above.

Source maps, such as climatic maps emphasizing temperature, relative humidity, and seasonality, and water maps showing pH and temperature, will have to be obtained or prepared. These may be especially important in revealing vector habitat locations. Landscape and relief maps, and vegetation maps, may also point out existing and potential breeding sites, etc., for the various vectors.

Quantification of Health Impacts

Available information on incidence and prevalence for the water-related disease under study in each project area and for the country as a whole should be collected. Incidence, which is a measure of the number of new cases of a disease reported in a given period of time (usually a year), is a good indicator for acute diseases such as cholera, infectious hepatitis, or typhoid. However, most existing incidence information is of varying degrees of poor quality, and the analysts may have to produce their own estimates. Prevalence, which is a measure of the total number of cases of a disease at any particular time, is a better measure for chronic diseases such as schistosomiasis. One can usually get better prevalence data than incidence data.

Official reports of disease-specific incidence or prevalence may either overestimate or underestimate the real situation, sometimes by orders of magnitude. With some exceptions, most diseases are underestimated (lack of coverage, reporting only from facilities, etc.). Malaria, in contrast, is often reported on the basis of fever alone (presumptive diagnosis), a practice which may lead to overreporting. The incidence or prevalence data should be sought according to the smallest available administrative unit (polygon) on the maps to be utilized for the analysis.

But incidence and prevalence indicators do not fully capture the relative importance of diseases in terms of their burden on the human population. The number of cases is not related directly to the actual disease burden. For example, 5,000 cases of malaria is different from, and more important than, 5,000 cases of gastroenteritis. A composite indicator of health status, which captured not just the incidence and prevalence but also the temporary and permanent disability and the premature mortality associated with a disease, would express much better the disease burden, and permit easy and meaningful comparisons, particularly by analysts and decision-makers who lack medical training. The most commonly used among such composite indicators is the number of healthy days (or potential years) of life lost.[19]

In spite of all its advantages, serious theoretical problems were encountered when the Study consultants tried to assess the number of healthy days lost for the country studies. Four of them are the:

1. indicator is useful with static populations (i.e., when irrigation does not imply immigration); it is difficult to conceptualize it when a massive immigration would be associated with irrigation development;

---

19. A Quantitative Method of Assessing the Health Impact of Different Diseases in Less Developed Countries, by the Ghana Health Assessment Project Team, _International Journal Epidemiology_, Vol. 10, No. 1, pp. 73-80, or the Potential Years of Life Lost construct which appears monthly in the _Morbidity and Mortality Weekly Report_, published by the U.S. Public Health Service, Center for Disease Control.

2. average added value per day is used to quantify in economic terms the worth of those days lost; this factor is rather meaningless when unemployment and under-employment are high. For example, diseases striking most seriously in the idle seasons, when the opportunity cost of labor is zero or near zero, would show up with a very low economic worth;

3. computation and valuation of the healthy days lost when a child dies is complex and, most likely, meaningless, and

4. computation of the worth of shifts in the days lost between different age groups is almost impossible and most likely meaningless [the example which led to this problem is the introduction of permanent irrigation in an area where seasonal malaria prevails. In the with project situation malaria may become endemic: child mobility and mortality would increase, while the surviving adults' mobility and mortality would decrease, since these would develop some kind of resistance].

More theoretical work is needed before an effective composite indicator is available for assessing the health risks of irrigation development.

Health impacts (however quantified) need to be examined in terms of the existing situation and its trend (the without-project situation), as well as for several possible scenarios of irrigation development (with-project situations). One of the major consequences of irrigation projects is massive immigration; population may double, triple, etc., simply because of the existence of the project. Therefore, the impact will have to be predicted on the basis of several possible population levels. It is suggested that twofold, threefold, fivefold, and tenfold increments from the base line population be utilized to estimate the disease burden. These scenarios will relate later to the needs for health education and health services as well as for other project-related investments.

Actions Recommended

Some safeguards can be built into a project's design to reduce, minimize, or pre-empt the anticipated health risks (see Table 2). Safeguarded projects may cost more than projects without safeguards: some safeguards may increase the project's capital costs; others, the recurrent costs after project completion. Therefore, it is necessary to assess the detrimental effect of the disease risks and compare them with the cost of the possible safeguards, so as to make explicit both the safeguards that thereby become justifiable and the ones that do not.[20]

---

20. For an excellent malaria-related study from India, relevant from both the health and economic viewpoints, see Jayaraman, T.K., Malarial Impact of Surface Irrigation Projects: A Case Study from Gujarat, India, Agriculture and Environment, 7 (1982), pp. 23-34.

Safeguards that involve increased recurrent expenditure after completion are more difficult to manage and to finance than those which involve increased initial capital costs. Further, they continuously create opportunities for something to go wrong, to not be done, or to be done inadequately. Therefore, their utility must be assessed within the context of how things actually work in the country.

Safeguards may fall into the following six categories:

1. <u>Project design</u>. Such things as the lining of canals, the "verticalness" of canal walls, the canal slope, the speed with which the water flows, the presence of siphons, and whether these siphons are covered or open, have an impact on the potential for water-related disease transmission as well as on the cost of the project. The silting of villages clear of a schistosomiasis-prone irrigated area has been well-documented as an effective way in reducing transmission[21] and can supplement the additional control structures that engineering can create.

2. <u>Project operation</u>. In some cases, changes or periodical fluctuations in water levels decrease the suitability of habitat for mosquitos, snails, or flies. Usually, such changes or fluctuations require some particular design features in the basic infrastructure, too.

3. <u>Supplementary investments</u>. Drinking water or laundering facilities supplied to the people who will reside in the project area may deter them from going to the canals and hauling home irrigation water by the bucket load. This reduces the number and length of contacts of local inhabitants with contaminated water, and thus reduces the incidence of some diseases. The construction of holding tanks, and fences (to decrease livestock or child water contract), or changing or adding to the project in other ways also might decrease the negative health impact.

4. <u>Health services and social services</u>. These should be provided in the project area to take care of the normal disease burden of the local population, as well as to deal effectively with whatever disease may result from the project itself.

5. <u>Health education</u>. Effective health education to modify populationbehaviors with negative health impacts might reduce risk. It should be a part of any project. In many cases, such health education should be started at the very onset of a project;

---

21. Sturrock, R.F., Bilharzia Transmission on a New Tanzanian Irrigation Scheme, <u>East African Medical Journal</u>, Vol. 43, No.1, Jan. 1966.

the ex-ante assessment of health risk would indicate the convenience of doing so.

6.   Epidemiological surveillance and monitoring. The bottom line of any safeguard to health must include an independent competence to measure the disease situation, to check (and confirm) pre-project expectations, and to see to it that any unexpected risks are identified early on with a direct channel for remedial or preventive action to be taken subsequently. Epidemiological surveillance is necessary to monitor the normal operation of the project, so that disease levels are assured to be within the limits set as tolerable by the people who accepted the project with its corresponding design specifications.

Many if no all of the actions to be recommended may require additional investments, but health costs are not usually a major factor in capital-intensive irrigation projects. However, it would be prudent to bear in mind that some of these costs can escalate beyond the point of endangering its economic viability.

Table 2

Preventive Strategies for Water-Related Diseases

Category	Preventive Strategy
Water-borne	Improve water quality. Prevent casual use of unimproved sources.
Water-washed	Improve water quality. Improve hygiene. Improve water accessibility.
Water-based	Decrease water contact. Control snails. Improve water quality.
Water-related Insect Vectors	Improve surface water management. Destroy breeding sites. Decrease human-insect contacts.

Source: Adapted from Feachem, R., McGarry, M., Mara, D., op. cit.

In practical terms, administrative capabilities and administrative arrangements are the "Achilles Heel" of irrigation projects. It is when the weeds do not get taken out, when erosion is not dealt, with, when pumps are not maintained, when seepage is allowed to occur and does not get drained off, that well-designed and well-constructed irrigation projects incur unanticipated and large-scale negative health impacts.

General Procedures

The Preparation Phase

Often, surprisingly good information can be quickly gathered outside the country, avoiding time-consuming searches for equivalent or lesser quality information in-country. Time spent in good preparation will pay off in the field. The main procedures for such search, as well as the main sources within the bank, are discussed at Annex 4.

Outside the Bank, other UN agencies such as WHO, UNDP, and FAO, must be consulted. At WHO headquarters, an atlas of the global distribution of schistosomiasis (to be published in collaboration with the Centre d'Etudes de Geographie Tropicale at Bordeaux) is now being developed. When available, it may be obtained from the Chief of the Schistosomiasis and Other Snail-Borne Trematode Infections, in the Parasitic Disease Program (currently, Dr.K. Mott).

Not to be missed at WHO is the Tropical Disease Research Group, as well as persons who may have specific information regarding health conditions for water-related diseases in particular countries. The WHO regional offices [22], and the WHO Programme Coordinator Offices (usually in each country) hold most of the country-specific relevant information. PEEM, the Panel of Experts on Environmental Management, has a very active committee for vector control (with many useful documents) and is reachable through WHO headquarters [23], though it has FAO and other representation composition.

Other good sources within the donor community are the regional development banks, and the various bilateral donor agencies such as USAID, the Danish, German, and Canadian International Development Agencies, and so on.

Lastly, institutions of academic excellence might well be consulted. The two great schools of tropical medicine in the U.K. (London and Liverpool) come to the fore, but excellent information also exists in Belgium, Holland, France, Germany, the U.S., and many other places. Especially helpful may be the cache of theses and dissertations, that often contain candid detail and contact information.

---

22. As PAHO for the Americas, EMRO for the Eastern Mediterranean, AFRO for much of Africa, EURO for Europe and Soviet Asia, SEARO for Southeast Asia, and WPRO for the Western Pacific and China.

23. Mr. R.J. Tonn, PEEM Secretary, VBC Division, WHO, 1211 Geneva, 27 Switzerland.

In-Country Work

The analysts will have relatively little time in the country to gather a great deal of information. One of the most important enabling things is to acquire a counterpart expert, who can facilitate information gathering and a proper understanding of what it is that one is seeing/hearing, as well as transportation arrangements and other arrangements that are so vital in yielding the needed information. Ideally, an officer from the Ministry of Planning or the Ministry of Health, who can spend some time travelling and who knows his or her own country well, would be the best facilitator.

Basically, the analysts are going to be working on two tracks: (i) gathering the specific information identified in the preparation steps outside the country, and (ii) some basic reconnaissance. The preparatory work may have provided them a list of potential sources for the specific information already identified as needed. They will have to confirm that those sources can actually deliver that information or whether better sources exist or need to be sought. Then, sources for the remaining information will have to be identified.

It is important to visit some local medical facilities in the national (or regional) capital and to see firsthand cases of the water-related diseases under study in order to get a feeling for how the diagnoses are made and the level of competence within the clinical arena. The national level of laboratory proficiency should also be assessed. For example, for cholera one needs specific culture media such as TCBS (thiocitrate bile salt) agar upon which the cholera organism can grow. Similarly, there is an enzyme-linked immunoassay for enteric viruses and a radioimmunoassay for hepatitis. For typhoid fever, one would want to know whether the diagnoses are made on clinical grounds or whether there is stool and/or blood culture with the actual organism frown for bacteriological confirmation.

It is also important to talk to the national experts in a few areas, particularly experts on the respective specific diseases. Experts on specific soils or vegetation which inhibit the growth of snails (or other vectors) may be of vital interest for irrigation with various plant species or with particular ecological environments in which the vectors either flourish or are discouraged from habitation. Especially important is to talk to people who represent the local variety of medicine, i.e., traditional medicine, and its capability to prevent or cure water-related diseases. These are largely untapped resources but are worth seeking out actively. Try to get a handle on the existing water and/or irrigation projects, and their management and logistics, especially weeding, vehicle maintenance, and machine maintenance. This will give some idea about the level of safeguards that should be built into a project to assure that the disease situation is going to be a prevented rather than a realized potential. It would also be important to see if the Ministry of Health is aware of the distribution of health services and social services in the rest of the country in order to match that existing level when one is planning services for the project area. Sometimes, an irrigation health hazard survey or other baseline study may have been conducted because a possible project has already been appreciated.[24]

Visiting some local areas, including some existing irrigation projects, is a vital part of getting a "feel" for the country. This will enable the analysts to see: (i) how they are managed, (ii) how much the people know and how they feel about water-related disease, (iii) whether they feel that malaria (or schistosomiasis or whatever) is a bigger problem than they had been led to believe, and (iv) whether they were promised different things before the project than were actually delivered after project completion. The analysts should also try to get a handle on what diseases are important and which ones are a source of concern for local residents, not from the bias of whatever quantitative epidemiological perspective a consultant brings, but from the people themselves. It would also be important to visit some health posts at random to check their functioning, to check their recording systems to define their reporting frequency, to see what happens when unusual occurrences are reported, to see how well the referral system works, and to see their logistic supply and transport situations.

One should always be on the lookout for behaviors that can amplify water-related disease, such s children playing in irrigation canals, or the drinking of canal water, or particular local habits of fecal hygiene which may play into the hands of a particular disease and facilitate its transmission. The following excerpt from an article on schistosomiasis associated with irrigation schemes in Ethiopia makes this point succinctly:

> "In all except one, the labor villages are provided with safe water from artisan wells. Many women prefer, however, to wash the clothes in the canals on or near the concrete constructions of the irrigation system, such s distribution works, culverts, etc. Here they have larger quantities of water at their disposal than in the villages where is a limited number of taps. Moreover, the flat concrete constructions lend themselves well for kneading and beating clothes or drying the wet clothes in the sun. Children who accompany their mothers and play in the water are also exposed".[25]

One should also be on the lookout for behavior that can decrease water-related disease, such as some plant or other substance which may be rubbed on skin by local villager to prevent schistosomiasis or mosquito bites.

---

24. For an excellent prototype, see Bunnag, T., et.al., Potential Health Hazards of the Water Resources Development: A Health Survey in the Phitsanulok Irrigation Project, Nan River Basin, Northern Thailand, Southeast Asian Journal of Tropical Medicine Health, and Public Health Vol. 11, No. 4, Dec. 1980.

25. Bruijning, CFA. Bilharziasis in Irrigation Schemes in Ethiopia, Tropical and Geographic Medicine, 21 (1969), 280-292.

Synthesis: Towards a "Diagnosis"

The most difficult part of this methodology must now be addressed: the logic that will tie all the assorted bits of information collected into a rational and unified assessment of the risks to human health from adding an irrigation project to any given landscape. This is the logic that gives a rationale to, and defines, the information which should be collected; the thought process of making a diagnosis.

The Chain of Reasoning

The first evidence to be weighed must be whether or not significant irrigation-related disease is present in the country. Then, the diseases selected for the study must be addressed, along with vector prevalence, potential for vector habitat creation/extension, and the reliability of this information.

The areas where the disease exists at an endemic or hyper-endemic level would be labelled as areas already infected. If disease exists at the hypo-endemic level or with only occasional outbreaks, climatic, vegetation, and other ecological data should be then considered to assess whether there is a significant risk that the increase and changes in patterns or availability of water and of population might create the conditions which may lead to epidemic or hyper-endemic situations. If the disease does not exist but the vector does, the project area may be labelled as an area without disease that could become highly receptive to the disease after irrigation development; drastic control and monitoring safeguards should be built into the project. If neither the disease nor the vector exist, an analysis of the climatic, vegetation, and other ecological data should provide information on whether the changes in the quantity or distribution of water and/or population to be brought about by irrigation would create conditions for the introduction of the vector and/or the disease. According to the level of probability assessed by the analysts, the area could be labelled as currently without disease that could become either highly receptive or marginally receptive to disease after irrigation is introduced. If neither the disease nor the vector currently exist, and if neither climatic nor other ecological conditions would favor their development, the area could then be labelled as a low risk or risk-free area. In short, the analysts should look first into the pre-existence of the disease or the vector, then at the climatic and other ecological conditions which might indicate whether conditions permit or prevent the development of either, and then at the modifications in the water and population density, amount or distribution which the project is meant or likely to originate.

The Actual Procedure

This general chain of reasoning is a description of the logical steps which guide the analysts' work. It is also the procedure to be actually followed when the task of the analysts is to analyze the health risks associated with one individual project. However, the actual procedure changes when the subject is the health risks associated with irrigation development in a whole region or in the country at large. In this case, maps should be prepared showing the current distribution of each of the diseases under study and their vectors, as well as maps showing the areas where climatic or other ecological conditions favor, are neutral to, or retard or prevent the development of either the disease or the vector. The two sets of maps are overlaid for each of the diseases. The areas where each of the diseases exist and climatic and other ecological conditions favor its establishment or development are labelled as areas already infected with high levels of transmission.

Similarly, the areas where neither the disease nor the vector currently exist and where other conditions do not favor their development could be labelled as risk-free areas. The areas where the disease exists and other conditions are neutral or not favorable would require additional analysis. The kind of irrigation project to be established should be analyzed to check whether the addition of water or of population in the amounts, density and distribution which are characteristic of that kind of project could lead to an epidemic or hyper-epidemic condition, or whether the current pattern of disease resistance will continue. A similar analysis should be conducted where the disease does not exist but the vector is already established.

The areas with favorable conditions and where the disease does not exist can also be labelled as areas with risk of becoming receptive to the disease after irrigation development, while the areas without the disease and with climatic and other ecological conditions neutral could be labelled as areas which could become marginally receptive after irrigation development. All analyses are carried out on an area basis. No analysis is done project-by-project except in those cases where an area may be grouped into either one or another sub-set depending on the way water is going to be distributed, population to be settled, etc. The end result is a map of "disease-proneness", where the whole country or region is divided into any potential for disease is identified, this must be judged in the context of the country and the practicality and cost-effectiveness of the possible safeguards. The result is a judgment, a considered opinion of the risk to human health from a specific irrigation project.

Chapter 18

DAMS AND THE ENVIRONMENT: ECONOMIC ASPECTS[26]

John A. Dixon

Dams are large, lumpy social investments made by societies to provide social benefits. As such it is appropriate that the economic analysis of dam projects be carried out on both financial and social welfare criteria. This, in fact, is normally done since many of the benefits produced by dam projects (e.g. flood control, irrigation, recreation) are difficult to capture directly. Other benefits, such as power and domestic/industrial water, can be sold and valued much more easily.

A problem arises, however, when environmental, social, or resource effects of dams are not bought or sold an dare not valued in any market. These impacts should still form part of the total benefit-cost analysis of a dam project, and yet they are frequently excluded. This exclusion, and the magnitude of some of the impacts, are one reason why many large dam projects are criticized and generate considerable worldwide attention.

In order to assist in the expanded analysis of the social welfare (or economic) evaluation of dam projects, the Bank's Agriculture and Rural Development Department has supported an examination of environmental considerations and their related economic aspects. This work is contained in a new report, *Dams and the Environment: Considerations in World Bank Projects* (Dixon, Talbot, Le Moigne, 1989).

A Matter of Benefits and Costs

Much of the discussion at the seminar today has focussed on the impacts of dams on the environment, and of the environment, in turn, on dams. The economist must start with an understanding of these impacts and needs to work with engineers, natural scientists and other social scientists to do so.

Both direct and indirect effects must be considered within a with-project and without-project format. (In addition, the question of welfare impacts of alternative water or power development projects, if the dam is not built, may also need to be considered.)

---

26. This extended abstract is drawn from Chapters 4 and 5 of Dams and the Environment: Considerations in World Bank Projects by J.A. Dixon, L.M. Talbot and G.J.M. Le Moigne, forthcoming, World Bank, 1989.

Once environmental and/or social impacts are identified, they need to be expressed in monetary terms, to the extent possible. This is the process of <u>valuation</u> and considerable work has been carried out on this area in recent years. Table 4.1 from our report present selected environmental effects, their economic impact, both benefits and costs, and representative valuation techniques. Some of these techniques are outlined in Chapter 4, and given in greater detail in the literature cited in the text.

A broader social analysis should help to design better projects by assessing, in a comprehensive manner, the benefits and costs associated with a proposed project. This will not, however, remove all controversy. There will remain certain areas that are not amenable to economic analysis.

Table 4.1 Selected Environmental Effects and Their Economic Impacts

Environmental Effect	Economic Impact	Benefit (B) Cost (C)	Representative Valuation Technique
Environment on Dams			
1) Soil Erosion - upstream, sedimentation in reservoir	reduced reservoir capacity; change in water quality	B,C	change in production, preventive expenditures, replacement costs
Dams on the Environment			
1) Chemical Water quality-changes reservoir and downstream	increased/reduced in treatment cost, reduced fish catch, loss of production	B,C	preventive expenditures, changes in production
2) Reduction in silt load - downstream	loss of fertilizer reduced siltation of canals, better water control	B,C	replacement costs, preventive expenditures avoided
3) Water Temperature changes (drop)	reduction of crop yields (esp. rice)	C	changes in production
4) Health - water related diseases	sickness, hospital care, death	B,C	loss of earnings, health care costs
5) Fishery - impacts on fish irrigation, spawning	both loss and increase in fish production	B,C	changes in production, preventive expenditures

Environmental Effect	Economic Impact	Benefit(B) Cost (C)	Representative Valuation Technique
6) Recreation - in the reservoir or river	value of recreation opportunities gained or lost, tourism	B,C	travel cost approach, property value approach
7) Wildlife and Biodiversity	creation or loss of species habitat and genetic resources	B,C	opportunity cost approach, tourism values lost, replacement cost
8) Involuntary Resettlement	cost of new infrastructure, social costs	C	replacement cost approach, "social costs" relocation costs
9) Discharge variations excessive diurnal variation	disturbs flora and fauna, human use	C	relocation costs, changes in production
10) Flood attenuation reduces flood damage	reduces after flood	B,C	changes in cultivation; production, flood damages avoided

For example, cultural or spiritual values are not easily valued. Similarly, determining the monetary benefits of genetic diversity is very difficult.

In addition, there are people who fundamentally distrust economists and any economic analysis (except, it seems, when it supports their predetermined position)! In the Goldsmith-Hildyard report on *The Social and Environmental Effect of Large Dams* (1985), for example, the following *sins* of economic analysis (and economists) are listed:

1. Using unrealistically low discount rates,
2. Over-estimating job creation potential,
3. Failing to account for energy costs,
4. Over-estimating benefits of flood control,
5. Ignoring costs of decommissioning,

6.   Over-estimating the life of dams,
7.   Under-estimating construction cost,
8.   Over-estimating irrigation and recreation benefits, and
9.   Under-estimating environmental costs.

Most of these criticisms can be met by a competent, careful economic analysis, especially when sensitivity analysis is applied to key variables. Other criticisms, such as the first one about discount rates, reflect differing philosophies and are not easy to resolve. The last criticism, under-estimating environmental costs, may have some merit and in fact, if true, is reflected in over-estimation of many anticipated benefits.

Economic analysis will not resolve the controversies surrounding large dams. Carefully done economic analysis may help, however, to clarify the real differences between proponents and opponents and sometimes produces unexpected results. For example, recent work on the benefits of watershed management measures above dams indicate that benefits are considerably smaller than expected. A better understanding of erosion and sediment transport processes will assist dam engineers, and economic analysts, to develop cost-effective measures to control sedimentation. A recent World Bank Technical Paper by Mahmood (1987) provides valuable insights on this process.

Another major problem area remains. When one probes the real reason for opposition to large dams and the ecological, environmental, and economic arguments used, one frequently finds that the underlying tension is a social-political one, particularly when resettlement issues are important. These issues are poorly handled by engineers or economists. This problem highlights again the need for communication, and professional modesty. (Cernea's recent Technical Paper on involuntary resettlement, 1988,clearly presents the main issues involved).

Selected Examples

The Dams and the Environment report contains a number of examples of major storage dams and their associated environmental factors. In many cases potential environmental problems did not materialize or were mitigated; in other cases, unanticipated problems arose. In only a few of the completed projects were wide ranging economic analyses carried out; environmental problems were usually described qualitatively.

The cases that are briefly described in the report include such major projects as the Tarbela Dam in Pakistan and the Aswan High Dam in Egypt. These projects have been in existence long enough for ex-post evaluations to be carried out. Most environmental effects had been anticipated but sometimes their impacts, both positive and negative, were considerably different in magnitude than what was projected.

The Itaipu Dam on the Brazil/Paraguay border is a recently completed project; resettlement issues have been the biggest social and political problem. Two other projects are either under construction (the Narmada Valley Project) or under consideration (Three Gorges). In both cases, opponents are very active and have focused on resettlement and ecological issues. The last case described, the Nam Choan project in Thailand, has been indefinitely postponed, largely due to environmental concerns.

The new pressures from both outside and within the Bank to examine the wider range of environmental impacts (and their economic implications) should produce very interesting analyses of upcoming projects. Resettlement costs, for example, are a major concern in discussion of the Three Gorges project; loss of habitat and wildlife figure large in many Amazonian project. Seminars such as this one are a valuable aid in meeting this goal.

ANNEX

LIST OF PARTICIPANTS
AND
REFERENCES

WORLD BANK

SEMINAR ON DAM SAFETY AND ENVIRONMENT

April 14-15, 1989

List of Participants

Country Representation

BRAZIL
Mr. Flavio Lyra — ICOLD Past President, President, Enge-Rio Engenharia E Consultoria -Consulting Firm

CANADA
Dr. Thomas Vladut — RETOM Geo-Research, Engineering Consulting Firm

FRANCE
Mr. D. Bonazzi — Coyne et Bellier, Consulting Firm
Mr. Jean Billore — Coyne et Bellier, Consulting Firm

INDIA
Y.K. Murthy — World Bank Consultant

UNITED KINGDOM
Mr. Paul A. Back — Sir Alexander Gibb, Consulting Firm
Mr. Robert Rangeley — ICID Past President, World Bank Consultant

UNITED STATES
Mr. Jan Veltrop — President ICOLD, Senior Vice President, Harza Consultants

Mr. Lloyd A. Duscha — U.S. Army Corps of Engineers

Mr. Weatherly — NGO Outreach Projects

Mr. Montague Yudelman — World Bank Consultant

Mr. T. W. Mermel — World Bank Consultant

Mr. Jack L. Davis — Tennessee Valley Authority

Mr. John A. Dixon — Environment and Policy Institute East-West Center, Honolulu

AGENCIES

INTER-AMERICAN DEVELOPMENT BANK
Mr. Blandon Upeguy

U.S. AGENCY FOR INTERNATIONAL DEVELOPMENT (USAID)
Mr. W. Fitzgerald

CANADA INTERNATIONAL DEVELOPMENT AID (CIDA)
Mr. A. Shady

WORLD BANK STAFF

SECTOR POLICY AND RESEARCH (PRE)
Mr. V. Rajagopalan, Vice President

Agriculture and Rural Development Department
Messrs. Le Moigne, Ochs, Plusquellec

Environment Department
Mr. J. Warford

Infrastructure and Urban Development Department
Mr. Adtavala

AFRICA REGION
Messrs.   Rama Skelton, AF1IE; Aizad Khan, AF2AG;
                 Patricio Millan, AF31E; Denis Robert, AF4IN;
                 E. Aikins-Afful, Anders G. Zeijlon, AF6IN;
                 A. Elahi, AFTAG; Leif Christoffersen, AFTEN;
                 Bocar Thiam, Said Mikhail, AFTIE;
                 Randolph Andersen, AFTIN

ASIA REGION
Messrs./Mdms.   Donal T. O'Leary, AS1IE; Israel Naor, AS2AG;
                 Daniel Gunaratnam, Victoria Elliott, AS3AG;
                 Chris Perry, Christoph Diewald, AS4AG;
                 Etienne Linard, Alfonso Sanchez, AS4TE;

Tom Blinkhorn, AS4CO; Vatsal Thakor, AS5IE;
Brian Albinson, Jagdish Srivastava, ASTAG;
Robert Morton, Barry Trembath, ASTEG;
Colin Rees, William Partridge,
Narendra Sharma; ASTEN

EMENA
Messrs.   H. Arima, EM1EG; J. Pierre Villaret,
Carlos Gois, EM2AG; Saeed A. Rana, EM3AG;
C.K. Chandran, EM41E; Luis Moscoso, EMTAG;
Spyros Margetis, EMTEN; David Howarth, EMTIN

LATIN AMERICAN REGION
Messrs.   Mohan Munasinghe, Emilio Rodriguez, LA1IE;
Jean-Louis Ginnsz, LA1AG; Jacques Martinod,
LA3AG; Peter Wittenberg, LA4AG;
Luis Luzuriaga, LA2IE; Armando Araujo,
Emanuel Idelovitch, LA4IE; Robert Goodland,
D. Crompton-Calvo, LATEN; Tim Campbell,
Rafael A. Moscote, Guillermo Yepes, LATIE

WATER AND SANITATION DIVISION
Mr. Alfonso, INUWS

LEGAL OPERATIONS
Ms. Barbara Lausche, LEGOP

References

Cernea, M.
  <u>Involuntary Resettlement in Development Project</u>. World Bank Technical
  Paper No. 80, 1988, Washington, D.C., The World Bank.

Dixon, J.A., L.M. Talbot and G. J-M. Le Moigne
  <u>Dams and the Environment: Considerations in World Bank Projects</u>, 1989,
  Washington, D.C., The World Bank.

Goldsmith, E. and N. Hildyard
  <u>The Social and Environmental Effect of Large Dams</u>, 1985, Wadebridge
  Ecological Center, United Kingdom

Mahmood, K.
  <u>Reservoir Sedimentation: Impact, Extent, Mitigation</u>. World Bank Technical
  Paper No. 71, 1987, Washington, D.C., The World Bank.

Distributors of World Bank Publications

ARGENTINA
Carlos Hirsch, SRL
Galeria Guemes
Florida 165, 4th Floor-Ofc. 453/465
1333 Buenos Aires

**AUSTRALIA, PAPUA NEW GUINEA,
FIJI, SOLOMON ISLANDS,
VANUATU, AND WESTERN SAMOA**
D.A. Books & Journals
11-13 Station Street
Mitcham 3132
Victoria

AUSTRIA
Gerold and Co.
Graben 31
A-1011 Wien

BAHRAIN
Bahrain Research and Consultancy
    Associates Ltd.
P.O. Box 22103
Manama Town 317

BANGLADESH
Micro Industries Development
    Assistance Society (MIDAS)
House 5, Road 16
Dhanmondi R/Area
Dhaka 1209

   Branch office:
   156, Nur Ahmed Sarak
   Chittagong 4000

BELGIUM
Publications des Nations Unies
Av. du Roi 202
1060 Brussels

BRAZIL
Publicacoes Tecnicas Internacionais
    Ltda.
Rua Peixoto Gomide, 209
01409 Sao Paulo, SP

CANADA
Le Diffuseur
C.P. 85, 1501B rue Ampère
Boucherville, Quebec
J4B 5E6

CHINA
China Financial & Economic Publishing
    House
8, Da Fo Si Dong Jie
Beijing

COLOMBIA
Enlace Ltda.
Apartado Aereo 34270
Bogota D.E.

COSTA RICA
Libreria Trejos
Calle 11-13
Av. Fernandez Guell
San Jose

COTE D'IVOIRE
Centre d'Edition et de Diffusion
    Africaines (CEDA)
04 B.P. 541
Abidjan 04 Plateau

CYPRUS
MEMRB Information Services
P.O. Box 2098
Nicosia

DENMARK
SamfundsLitteratur
Rosenoerns Allé 11
DK-1970 Frederiksberg C

DOMINICAN REPUBLIC
Editora Taller, C. por A.
Restauracion e Isabel la Catolica 309
Apartado Postal 2190
Santo Domingo

EL SALVADOR
Fusades
Avenida Manuel Enrique Araujo #3530
Edificio SISA, ler. Piso
San Salvador

EGYPT, ARAB REPUBLIC OF
Al Ahram
Al Galaa Street
Cairo

The Middle East Observer
8 Chawarbi Street
Cairo

FINLAND
Akateeminen Kirjakauppa
P.O. Box 128
SF-00101
Helsinki 10

FRANCE
World Bank Publications
66, avenue d'Iéna
75116 Paris

GERMANY, FEDERAL REPUBLIC OF
UNO-Verlag
Poppelsdorfer Allee 55
D-5300 Bonn 1

GREECE
KEME
24, Ippodamou Street Platia Plastiras
Athens-11635

GUATEMALA
Librerias Piedra Santa
Centro Cultural Piedra Santa
11 calle 6-50 zona 1
Guatemala City

HONG KONG, MACAO
Asia 2000 Ltd.
6 Fl., 146 Prince Edward Road, W.
Kowloon
Hong Kong

HUNGARY
Kultura
P.O. Box 139
1389 Budapest 62

INDIA
Allied Publishers Private Ltd.
751 Mount Road
Madras - 600 002

   Branch offices:
   15 J.N. Heredia Marg
   Ballard Estate
   Bombay - 400 038

   13/14 Asaf Ali Road
   New Delhi - 110 002

   17 Chittaranjan Avenue
   Calcutta - 700 072

   Jayadeva Hostel Building
   5th Main Road Gandhinagar
   Bangalore - 560 009

   3-5-1129 Kachiguda Cross Road
   Hyderabad - 500 027

   Prarthana Flats, 2nd Floor
   Near Thakore Baug, Navrangpura
   Ahmedabad - 380 009

   Patiala House
   16-A Ashok Marg
   Lucknow - 226 001

INDONESIA
Pt. Indira Limited
Jl. Sam Ratulangi 37
P.O. Box 181
Jakarta Pusat

IRELAND
TDC Publishers
12 North Frederick Street
Dublin 1

ITALY
Licosa Commissionaria Sansoni SPA
Via Benedetto Fortini, 120/10
Casella Postale 552
50125 Florence

JAPAN
Eastern Book Service
37-3, Hongo 3-Chome, Bunkyo-ku 113
Tokyo

KENYA
Africa Book Service (E.A.) Ltd.
P.O. Box 45245
Nairobi

KOREA, REPUBLIC OF
Pan Korea Book Corporation
P.O. Box 101, Kwangwhamun
Seoul

KUWAIT
MEMRB Information Services
P.O. Box 5465

MALAYSIA
University of Malaya Cooperative
    Bookshop, Limited
P.O. Box 1127, Jalan Pantai Baru
Kuala Lumpur

MEXICO
INFOTEC
Apartado Postal 22-860
14060 Tlalpan, Mexico D.F.

MOROCCO
Societe d'Etudes Marketing Marocaine
12 rue Mozart, Bd. d'Anfa
Casablanca

NETHERLANDS
InOr-Publikaties b.v.
P.O. Box 14
7240 BA Lochem

NEW ZEALAND
Hills Library and Information Service
Private Bag
New Market
Auckland

NIGERIA
University Press Limited
Three Crowns Building Jericho
Private Mail Bag 5095
Ibadan

NORWAY
Narvesen Information Center
Bertrand Narvesens vei 2
P.O. Box 6125 Etterstad
N-0602 Oslo 6

OMAN
MEMRB Information Services
P.O. Box 1613, Seeb Airport
Muscat

PAKISTAN
Mirza Book Agency
65, Shahrah-e-Quaid-e-Azam
P.O. Box No. 729
Lahore 3

PERU
Editorial Desarrollo SA
Apartado 3824
Lima

PHILIPPINES
National Book Store
701 Rizal Avenue
P.O. Box 1934
Metro Manila

POLAND
ORPAN
Patac Kultury i Nauki
00-901 Warszawa

PORTUGAL
Livraria Portugal
Rua Do Carmo 70-74
1200 Lisbon

SAUDI ARABIA, QATAR
Jarir Book Store
P.O. Box 3196
Riyadh 11471

MEMRB Information Services
    Branch offices:
    Al Alsa Street
    Al Dahna Center
    First Floor
    P.O. Box 7188
    Riyadh

    Haji Abdullah Alireza Building
    King Khaled Street
    P.O. Box 3969
    Damman

    33, Mohammed Hassan Awad Street
    P.O. Box 5978
    Jeddah

**SINGAPORE, TAIWAN, MYANMAR,
BRUNEI**
Information Publications
    Private, Ltd.
02-06 1st Fl., Pei-Fu Industrial
    Bldg.
24 New Industrial Road
Singapore 1953

SOUTH AFRICA, BOTSWANA
For single titles:
Oxford University Press Southern
    Africa
P.O. Box 1141
Cape Town 8000

For subscription orders:
International Subscription Service
P.O. Box 41095
Craighall
Johannesburg 2024

SPAIN
Mundi-Prensa Libros, S.A.
Castello 37
28001 Madrid

Librería Internacional AEDOS
Consell de Cent, 391
08009 Barcelona

SRI LANKA AND THE MALDIVES
Lake House Bookshop
P.O. Box 244
100, Sir Chittampalam A. Gardiner
    Mawatha
Colombo 2

SWEDEN
For single titles:
Fritzes Fackboksforetaget
Regeringsgatan 12, Box 16356
S-103 27 Stockholm

For subscription orders:
Wennergren-Williams AB
Box 30004
S-104 25 Stockholm

SWITZERLAND
For single titles:
Librairie Payot
6, rue Grenus
Case postal 381
CH 1211 Geneva 11

For subscription orders:
Librairie Payot
Service des Abonnements
Case postal 3312
CH 1002 Lausanne

TANZANIA
Oxford University Press
P.O. Box 5299
Dar es Salaam

THAILAND
Central Department Store
306 Silom Road
Bangkok

**TRINIDAD & TOBAGO, ANTIGUA
BARBUDA, BARBADOS,
DOMINICA, GRENADA, GUYANA,
JAMAICA, MONTSERRAT, ST.
KITTS & NEVIS, ST. LUCIA,
ST. VINCENT & GRENADINES**
Systematics Studies Unit
#9 Watts Street
Curepe
Trinidad, West Indies

TURKEY
Haset Kitapevi, A.S.
Istiklal Caddesi No. 469
Beyoglu
Istanbul

UGANDA
Uganda Bookshop
P.O. Box 7145
Kampala

UNITED ARAB EMIRATES
MEMRB Gulf Co.
P.O. Box 6097
Sharjah

UNITED KINGDOM
Microinfo Ltd.
P.O. Box 3
Alton, Hampshire GU34 2PG
England

URUGUAY
Instituto Nacional del Libro
San Jose 1116
Montevideo

VENEZUELA
Libreria del Este
Aptdo. 60.337
Caracas 1060-A

YUGOSLAVIA
Jugoslovenska Knjiga
YU-11000 Belgrade Trg Republike